Environmental Footprints and Eco-design of Products and Processes

Series Editor

Subramanian Senthilkannan Muthu, Green Story Inc.
Toronto, Canada

Indexed by Scopus

This series aims to broadly cover all the aspects related to environmental assessment of products, development of environmental and ecological indicators and eco-design of various products and processes. Below are the areas fall under the aims and scope of this series, but not limited to: Environmental Life Cycle Assessment; Social Life Cycle Assessment; Organizational and Product Carbon Footprints; Ecological, Energy and Water Footprints; Life cycle costing; Environmental and sustainable indicators; Environmental impact assessment methods and tools; Eco-design (sustainable design) aspects and tools; Biodegradation studies; Recycling; Solid waste management; Environmental and social audits; Green Purchasing and tools; Product environmental footprints; Environmental management standards and regulations; Eco-labels; Green Claims and green washing; Assessment of sustainability aspects.

Amilton José Vieira de Arruda
Felipe Luis Palombini
Editors

Biology, Biomimetics and Natural Design

Innovative Technologies and Sustainable Materials

Springer

Editors
Amilton José Vieira de Arruda
BioDesign Lab, Programa de
Pós-Graduação em Design (PPGDesign)
Universidade Federal de
Pernambuco [UFPE]
Recife, Brazil

Felipe Luis Palombini
Department of Industrial Design
Universidade Federal de Santa Maria
Santa Maria, Brazil

ISSN 2345-7651 ISSN 2345-766X (electronic)
Environmental Footprints and Eco-design of Products and Processes
ISBN 978-3-031-65876-1 ISBN 978-3-031-65877-8 (eBook)
https://doi.org/10.1007/978-3-031-65877-8

This Springer imprint is published by the registered company Springer Nature Switzerland AG
The registered company address is: Gewerbestrasse 11, 6330 Cham, Switzerland

Contents

Chapter 1
Adaptive Façades: A Comparative Analysis of Bimetal Solutions

Tarciana Andrade, José Beirão, and Amilton José Vieira de Arruda

Abstract The current knowledge demonstrates that the expansion of adaptive façades, which contribute to improving the energy performance of built spaces, can combine biomimetics research into advanced materials. The bimetal is a smart material that bends when heated. Designers use the bimetal to propose architectural solutions adapted to different climatic conditions, classifying them, for example, into the following typologies: self-shading [SH], self-ventilating, self-energizing, and self-structuring. This work seeks to demonstrate the potential of bimetal in architecture. We report on 20 solutions, including installations and proposals for façades, published between 2010 and 2022, then classify them among the typologies mentioned before. We identified five solutions by bioinspiration. Modules make up the solutions. Together, they provide complex structures and geometries that are lightweight and responsive. We selected four SH proposals for comparative analysis because these typologies correspond to the scope of this research. SH proposals are designs of smart systems that use solar energy to block the sun, avoiding excess heat gain. The projects selected are modular solutions that can have different characteristics, such as being a commercial solution, aiming to promote visual permeability, containing a photovoltaic cell, or using bimetal as an actuator. We conclude that bimetal façades have the potential to provide solutions that combine creativity with climate performance, with unique standard elements. By associating nature's strategies with development, it is possible to develop responsive solutions through the movement of modules. The elements adapt to different environmental conditions (that affect air temperature) and improve the energy performance of buildings.

T. Andrade (✉) · J. Beirão
CIAUD, Research Centre for Architecture, Urbanism and Design, Lisbon School of Architecture, Universidade de Lisboa, Lisboa, Portugal
e-mail: jnb@fa.ulisboa.pt

A. J. V. de. Arruda
BioDesign Lab, Programa de Pós-Graduação em Design (PPGDesign), Universidade Federal de Pernambuco [UFPE], Recife, Brazil
e-mail: amilton.arruda@ufpe.br

A. J. V. de Arruda, F. L. Palombini (eds.), *Biology, Biomimetics and Natural Design*, Environmental Footprints and Eco-design of Products and Processes,
https://doi.org/10.1007/978-3-031-65877-8_1

Keywords Kinetic façade · Smart material · Bimetal · Thermobimetal · Biomimetic adaptive façade · Adaptive shading · Self-shading

1.1 Introduction

Biomimetics applied to architecture, in turn, presents transdisciplinary approaches that seek to connect biologists to designers and technologists [4, 19, 20]. It stands out for the research on new materials, their performance, and respective production processes [12, 19]. An attentive look at the solutions of nature, which has evolved over 3.85 billion years, can contribute to revolutionizing products, processes, and life [5].

To promote kinetic façades adaptable to different climatic conditions, extant research shows the intensification of solutions using smart materials [7, 12, 25] that aim to reduce system maintenance and achieve zero energy options. Solar shading and shutters can contribute to energy reduction during the different seasons and present a feasible reduction in CO_2 consumption rates [11].

Bimetal or thermobimetal is a smart material [24] that tends to bend when heated [22, 24]. The explanation for such behavior lies in the composition of the bimetal, which comprises at least two layers of metal, permanently bonded and with different thermal expansion coefficients [10, 19, 21, 22, 24] (Fig. 1.1). The temperature variation will cause the passive layer to expand less than the active one, causing the material to bend [10, 18, 21].

According to Sack-Nielsen [22], a bimetal exhibits isotropic properties (equal and predictable behavior for both the width and height of the material). The behavior of bimetal can be enhanced or even prevented depending on its shape [24]. There are several advantages to incorporating the inherent characteristics of bimetal in architectural projects, such as predictability, precision, power consumption reduction, and low maintenance [13]. The material also presents reversible behavior to promote solar protection of windows at times of high heat gain [1–3, 6–8, 13–17, 22–25].

Nowadays, there are countless applications, and we can find bimetals practically everywhere [10, 21]. We find it in products that regulate and control temperatures, such as electric stoves, irons, hair dryers, and systems for protecting electrical circuits.

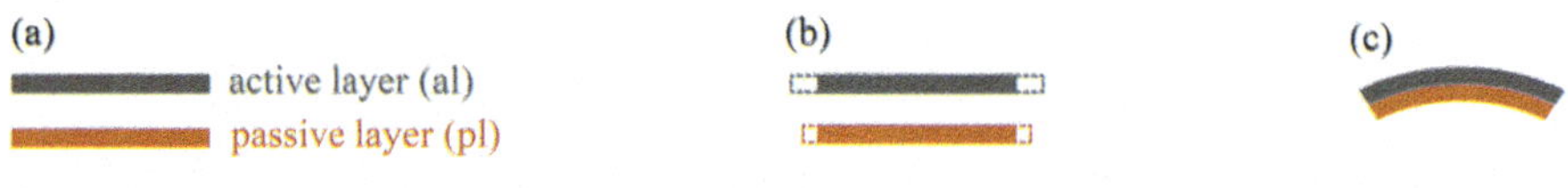

Fig. 1.1 Bimetal consists of two layers of metal with different rates of thermal expansion: (**a**) Layers that constitute the bimetal; (**b**) Each layer has different coefficients of thermal expansion; (**c**) Curvature of the material with the temperature rise: the active layer exerts compression, and the passive layer exerts tension. (Adapted from [2], License CC BY)

1.2 Bimetal in Architecture

In the last 20 years, thermobimetal has gained relevance in architecture [25]. Sung's work [9, 17, 24] is probably the most advanced example [24], relating form to performance [3, 24, 9]. She sought to develop solutions that are sensitive to environmental change, like human skin—without the use of energy and controls. Doris publishes the projects and experiments on the DOSU Studio Architects website [9] and classifies them into six typologies:

- ***Self-shading [SH]:*** Designs are smart systems that use solar energy to block the sun, avoiding excess heat gain, without the use of electromechanical devices.
- ***Self-ventilating [SV]:*** Projects that use the ambient air temperature to operate in order to release hot air. Once cooled, the system returns to its initial position.
- ***Self-structuring [ST]:*** Thicker versions of bimetal can be heated to higher temperatures (above 121 °C) to deform in the ideal condition for installation in tight spaces. Once cooled, the part can provide a pre-tensioned surface, resulting in a strong, light, and porous surface.
- ***Self-energizing [SE]:*** Uses solar energy in configuration to reflect, generate heat in a chimney, and provide air exchange, in order to ventilate the space.
- ***Self-assembling:*** Uses bimetal to facilitate toolless construction. In some cases, on non-reversible systems.
- ***Self-propelling:*** Experiments in the field of robotics to make things: spin, walk, roll, bounce, jump, slide, and shake. Such experiments use solar energy to drive movement and do not need computer controls or artificial intelligence to function.

Considering the first four typologies suggested by [9], we identified 20 bimetal proposals, published between 2010 and 2022, including installation [**IP**] and façade projects [**FA**] (Table 1.1). We also classified the proposals according to their inspiration in nature. We selected the first four typologies because of their relevance to this research, which aims to develop a self-shading solution for bio-inspired façades in the future. We classified the solutions as Complex Structures [**CS**] or Complex Movement [**CM**]. The classification process followed the definitions reported in the project descriptions.

We identified six projects inspired by nature. Tracheolis—ARS [**P5**] is bioinspired by the grasshopper's breathing system, which has a complex communication network between the vessels of the trachea. It is a proposal for self-ventilating. In turn, Urban Urchin [**P11**] emulates the morphology of the sea urchin. The aim of the project is to improve air quality through fluid dynamics. Air movement is associated with energy collection, a case of self-energizing. Solutions **P3**, **P18**, **P19**, and **P20** use the movement of plants as a reference. These proposals are examples of self-shading.

Ten solutions aimed to develop façade solutions [**FA**]. The other ten were installation proposals [**IP**]. *The self-shading typology corresponds to the classification with the highest incidence* (eleven proposals), followed by self-ventilating, self-energizing, and self-structuring (Graph 1.1).

Table 1.1 Twenty proposals with bimetal

Proposal	Year	FA or IP	B	SS or PM	C	Tipoloy SH	SV	ST	SE	Refs.
P1—Waist Tigtennig (DeCours, AIA New Orleans, LA)	2010	IP	N	S	CS	–	•	–	–	[9]
P2—Armoured Corset (WuHo Gallery, Hollywood, CA)	2010	IP	N	D	CS	–	•	–	–	[9]
P3—Facade application of an arrangement of bimetallic strips as a responsive Building envelope	2010	FA	Y	S	CS	•	–	–	–	[13]
P4—Bloom (M&Gallery, Silver Lake, CA)	2011	IP	N	D	CS	–	•	–	–	[9, 14, 24, 3]
P5—Tracheolis – ARS (Electronica, Linz, Austria) and (Copernicus Science Center, Warsaw, Poland)	2012 and 2015	IP	Y	D	–	–	•	–	–	[9]
P6—Blink Exhibition, MOCA; Los Angeles, CA	2013	IP	N	D	CS	–	•	–	–	[9]
P7—Bimetal Facade System DARC-Digital Architecture Research Consortium (Texas—EUA)	2013 – 2014	FA	N	S	CM	•	–	–	–	[15, 6]
P8—Pivot Shade System	2014	FA	N	S	CM	•	–	–	–	[24]
P9—Dialectic Form Finding of Passive and Adaptive Shading Enclosures	2014	FA	N	S	CM	•	–	–	–	[1]
P10—eXo (Museum of Contemporary Art, Santa Barbara, CA)	2014	IP	N	D	CS	–	–	•	–	[9]
P11—Urban Urchin (New York, NY I Chicago, IL I Los Angeles, CA)	2014	FA	Y	D	CS	–	–	–	•	[9]
P12—Tw/rl (Chicago, IL)	2015	IP	N	D	CS	–	–	–	•	[9]
P13—Shading system relies on the toggling of the shapes as the temperature changes	2016	FA	N	S	CM	•	–	–	–	[24, 9]
P14—Drift (New York, NY I Chicago, IL I Los Angeles, CA)	2016	IP	N	S	–	•	–	–	–	[9]
P15—Oculus	2017	IP	N	S	CM	•	–	–	–	[9]
P16—Performance through thickfolds (Danish)	2017	FA	N	S	CM	–	•	–	–	[22]
P17—Fuller 2.0 (South Coast Botanical Garden)	2018	IP	N	S	CS	–	–	•	–	[9]
P18—Invert Auto-Shading Windows	2019	FA	Y	S	CM	•	–	–	–	[16, 9, 23, 3]
P19—Pho'liage 01 (Brussels, Paris, and Luxembourg)	2019	FA	Y	S	–	•	–	–	–	[7, 17]
P20—Pho'liage 02 (Brussels, Paris, and Luxembourg)	2022	FA	Y	S	CM	•	–	–	–	[7]

Source: Authors

B = bio-inspired projects; **Y** = yes, meets the requirement; **N** = No, without classification; **FA** = Façade solutions; **IP** = Installation proposals. "•" = identifying the classification; "-" = the criterion in question did not occur. **S** = Single standard; **D** = Different modules; **CS** = Complex structures; **CM** = Complex movement

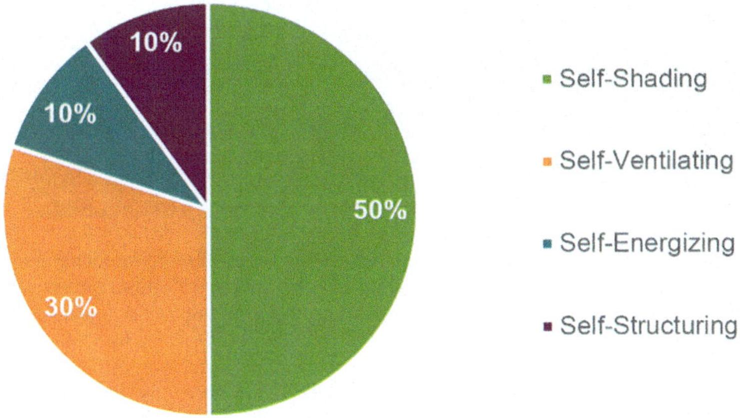

Graph 1.1 Typologies of the twenty bimetal projects

In the analysis, we emphasize that the classification as to typology (SH, SV, ST, or SE) should follow the project descriptions. Figure 1.2 shows different types of projects. However, we confirm that a solution could be multifunctional (such as the Urban Urchin project [**P11**], which also has a scope for self-shading and self-ventilating).

An example of a self-structuring solution is the eXo [**P10**]. The development process for this project consisted of preheating constituent bimetallic pieces and fitting them together in a pre-tensioned way. When the pieces cool down, they form a strong and lightweight structure, like the exoskeleton of a lobster or crustacean. The tension on the surface is maintained afterward.

Of the **FA**, eight were self-shading. *The studies focus on the performance of material movement to promote shading* (of the ten proposals, **CM** = 8; **CS** = 2). Nine solutions presented studies with a *single version of the element* (**S** = 9; **D** = 1), which allows for tessellation on the façade.

We identified the opposite for studies of installations that explore complex volumetrics (**CS** = 7; **CM** = 3); (**D** = 6; **S** = 4). However, we would like to point out that the installations are smaller-scale experiments, which could have the potential for façades, such as the Drift installation (**P14**).

Sixteen projects used *bimetal as the primary raw material* (**P1** to **P4**, **P6** to **P8**, **P10** to **P15**, and **P17** to **P19**). While others combined different materials with the solution, they used the *bimetal as the system's sensor and actuator* (**P5**, **P9**, **P16**, and **P20**). Therefore, two different development approaches allow the potential use of the material to be extended by combining the performance requirements of the façade with creativity.

Finally, this first analysis shows that several *solutions used parametric algorithms for their design*. These allowed for *structural complexity* (nine occurrences)

Fig. 1.2 (**a**) Bloom [P4], self-shading project; (**b**) Fuller 2.0 [P17], self-structuring solution; (**c**) Invert auto-shading windows [P18], self-shading project. Source: (**a**) photo by Brandon Shigeta (2011), CC-BY-SA-4.0, via Wikimedia Commons. Available at: https://commons.wikimedia.org/wiki/File:Bloom_by_DOSU.jpg (Accessed 12 October 2023); (**b**) photo by Doris Sung (2019), CC-BY-SA-4.0, via Wikimedia Commons. Available at: https://commons.wikimedia.org/wiki/File:Fuller_by_DOSU.jpg (Accessed 12 October 2023); (**c**) photo by Pinkelephante (2018), CC-BY-SA-4.0, via Wikimedia Commons. Available at: https://commons.wikimedia.org/wiki/File:Invert_Shading_System.png (Accessed 12 October 2023)

and a study of the *kinetic properties of standardized modules in bimetal* (eleven solutions). Thus, we can confirm that even though bimetal has isotropic characteristics, it is possible to create complex volumes or elements that can be adapted to different climatic conditions.

1.3 Comparative Analysis of Self-Shading Projects

The self-shading typology corresponds to the scope of this research. To better understand projects related to the SH, we selected four of them for comparative analysis (Table 1.2). As a selection criterion we defined that the project should necessarily: (a) be classified as self-shading; and (b) have a physical prototype. Thus, the following were selected:

- *InVert Auto-Shading Windows®*: A patented, awarded, and commercial product solution [3, 9, 16, 23].
- *Bimetal Facade System*: A version that combines the function of shading with the promotion of visual permeability [6, 15].
- *Pho'liage®* version 1: A prototype that uses bimetal associated with photovoltaic cells to convert excess solar energy into renewable energy [7, 17].
- *Pho'liage®* version 2: The bimetal in this prototype plays the role of an actuator, acting with the biopolymer (material responsible for promoting shading [7].

In the following, we briefly detail each of the selected projects that presents a comparative and qualitative analysis of the designs.

InVert Auto-Shading Windows features patents, awards, and a commercial product that reduces the need for artificial air conditioning by up to 25% [16]. The movement of plants inspires the design of the InVert™ modules, such as flowers or leaves that rotate to face the sun as it moves across the sky [16]. The modules sit inside the

Table 1.2 Comparative and qualitative analysis between designs using bimetal

Features	(a) Invert auto-shading	(b) Bimetal facade system	(c1) Pho'liage®	(c2) Pho'liage®
Research group or architecture studio	DOSU studio architects and TBM-designs LLC (Texas—USA)	DARC—Digital architecture research consortium (Texas—EUA)	ArtBuild (Brussels, Paris, and Luxembourg—EU)	ArtBuild (Brussels, Paris, and Luxembourg—EU)
Year	2019	2013–2014	2019	2022
Bioinspiration	•••	–	•••	•••
Self-shading	•••	•••	•••	•••
Visual permeability	••	•••	••	••
Modular solution	•••	•••	•••	•••
Number of bimetal parts per module	1	2	3	1
Bimetal thickness	0.06858 mm	0.09 mm	–	–
Prototype sizes	–	99.06 × 99.06 × 15.24 cm wood frame 50 bimetal louvers 6 vertical aluminums 30 connection pieces	–	–
System between glasses	•••	•••	•••	•••
Presence of photovoltaic cells	–	–	•••	–
Patent	•••	–	••	–
Award	•••	••	–	–
Commercial product	•••	–	–	–
References	[16, 9, 23, 3]	[15, 6]	[7, 17]	[7]

Source: Authors
We assigned "•", "••" or "•••" for each criterion defining a hierarchy of interest. The higher number of points means that more attention was given to the highlighted criterion. A "-" means that the criterion in question did not occur

cavity of a standard insulated glass unit (IGU) [16]. It acts as an organic shutter system that reduces heat gain and energy consumption [9]. The modules rotate under the effect of temperature variation and solar radiation [3] (Fig. 1.3).

The *Bimetal Facade System* seeks to promote a balance between the contradictory performance parameters of the façade: visual permeability and solar performance throughout the day or year [6, 15] (Fig. 1.4).

A prototype version uses bimetallic fins, 0.09 mm thick, fixed between a 3D printed connection and vertical aluminum rods [15] (Fig. 1.4a). It positions the rotated modular elements (Fig. 1.4b) around the fixing axis and allocates them between triple-glazed window systems. The bimetallic fins remain parallel and straight at room temperature. When heated they curve into an "S" shape and tend to be positioned more horizontally. By containing the fins between panes of glass they amplify thermal factors to activate the system [15]. The solution features shutters to manage the air masses between panes using fluid dynamics.

The first version of the *Pho'liage*® is a bioinspired design by the opening and closing of the *Oxalis acetosella* leaves that respond to light and temperature variation throughout the day and night (Fig. 1.5a). The solution is composed of three bimetal elements (Fig. 1.5b) and photovoltaic cells [7, 17] (Fig. 1.5c).

During the winter it aims to reduce the energy required for heating. In summer, it protects the façade from overheating. Monocrystalline photovoltaic cells (integrated into the solution) transform surplus solar energy into renewable energy [7, 17] (Fig. 1.5c). They performed an anticorrosion test with a painted mock-up positioned outdoors for 36 months.

The second version also uses the curve-line fold found on the leaf of *Dionaea muscipula* and *Aldrovanda vesiculosa* as bioinspiration (Fig. 1.6a, b). The solution features a tiny bimetal piece as an actuator. This piece is centrally attached to the triangular module with a curve-line fold, made of biopolymer—thermoplastic elastomers [7]. The bimetallic alloys function as actuators in the sun, while the biopolymer provides shading.

The responsive modules are fixed on adaptable honeycomb support structures made of aluminum. This solution corresponds to an adaptive shading device developed through biomimetics and simulation processes to optimize thermal comfort and contribute to reducing energy consumption [7].

Fig. 1.3 Invert auto-shading windows [P18], self-shading project. (Source: photo by Pinkelephante (2018), CC-BY-SA-4.0, via Wikimedia Commons. Available at: https://commons.wikimedia.org/wiki/File:Invert_Shading_System.png (Accessed 12 October 2023))

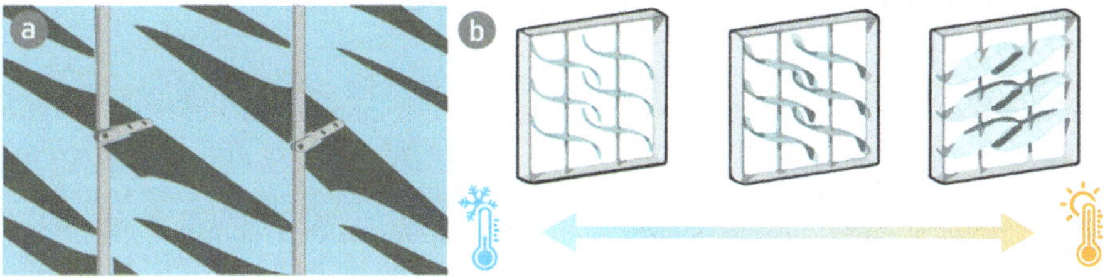

Fig. 1.4 Bimetal facade system. Authors based on the work of [6, 15]

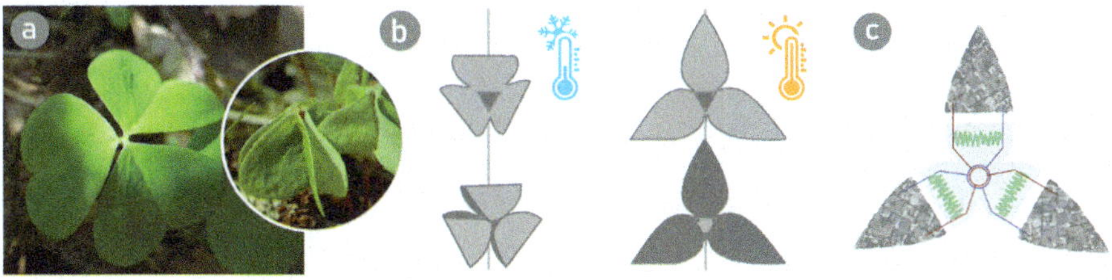

Fig. 1.5 (**a**) *Oxalis acetosella*; (**b**) three bimetal elements; (**c**) photovoltaic cells. (Source: (**a**) photo by Kristian Peters Fabelfroh (2005), CC-BY-SA-3.0, via Wikimedia Commons. Available at: https://commons.wikimedia.org/wiki/File:Oxalis_acetosella_blatt.jpeg (Accessed 12 October 2023); photo by Salicyna (2019), CC-BY-SA-4.0, via Wikimedia Commons. Available at: https://commons.wikimedia.org/wiki/File:Oxalis_acetosella_2019-04-04_8947.jpg (Accessed 12 October 2023); (**b**, **c**) Authors based on the work of [7, 17])

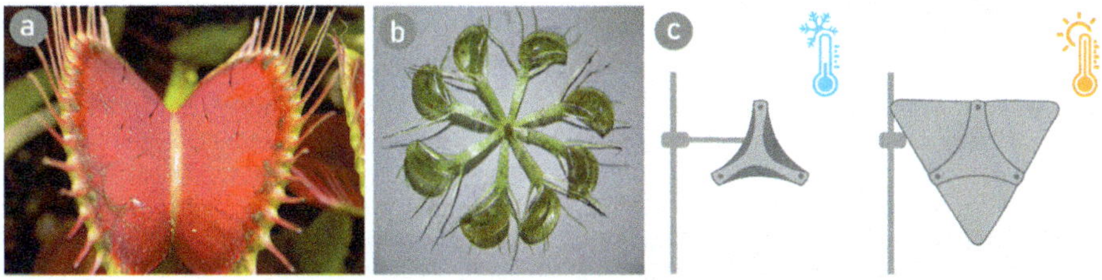

Fig. 1.6 Curve-line fold of: (**a**) *Dionaea muscipula*; (**b**) *Aldrovanda vesiculosa*; and (**c**) Module adaptable to different temperatures with the bimetal as an actuator. (Source: (**a**) photo by Noah Elhardt (2005), CC-BY-SA-2.5, via Wikimedia Commons. Available at: https://commons.wikimedia.org/wiki/File:Venus_Flytrap_showing_trigger_hairs.jpg (Accessed 12 October 2023); (**b**) photo by Stefan Lefnaer (2020), CC-BY-SA-4.0, via Wikimedia Commons. Available at: https://commons.wikimedia.org/wiki/File:Aldrovanda_vesiculosa_sl23.jpg (Accessed 12 October 2023); (**c**) Authors based on the work of [7])

1.4 Conclusions and Future Works

Architectural designs with bimetal can reduce the energy consumption of built spaces [7, 16]. This study sought to understand the *relevance of bimetal to architecture*. The analysis of the 20 projects showed that it is *possible to develop responsive and adaptable solutions, light and resistant structures, and complex geometries using bimetal*. In other words, creative approaches can go beyond the isotropic limits of the material.

The proposals for self-shading façades used modules between panes of glass and sought to improve the protection performance of the standard elements. *Concerning bimetal*, the analysis showed that we can use it for architectural projects as the *primary raw material or as a sensor/actuator in a system*. However, the composition of the bimetallic alloy, the thickness of the material, and the geometry of the façade devices could influence the kinetic capacity of the solutions requiring therefore specific research.

By combining nature's strategies with the intelligence of bimetal, we believe it is possible to provide adaptable solutions that are self-actuated by the weather and that work without electrical or mechanical devices. Our future publications will demonstrate this aspect of development.

Acknowledgments This work is financed by national funds through FCT—Fundação para a Ciência e a Tecnologia, I.P., under the Strategic Project with the references UIDB/04008/2020 and UIDP/04008/2020. FCT supports this study through grants SFRH/BD/144910/2019.

References

1. Adriaenssens S, Rhode-Barbarigos L, Kilian A, Baverel O, Charpentier V, Horner M, Buzatu D (2014) Dialectic form finding of passive and adaptive shading enclosures. Energies 7:5201–5220. https://doi.org/10.3390/en7085201
2. Andrade T, Beirão J, Arruda A, Cruz C (2021) The adaptive power of *Ammophila arenaria*: biomimetic study, systematic observation, parametric design, and experimental tests with bimetal. Polymers 13:1–17. https://doi.org/10.3390/polym13152554
3. Andrade T, Beirão J, Arruda A, Eysen C (2021) Toward adaptable and responsive facades: using strategies for transforming of the material and bio-based materials in favor of sustainability. In: Arruda A, Langella C, Bella D (eds) Cuaderno 149 | Centro de Estudios en Diseño y Comunicación (2021/2022), pp 37–59. https://doi.org/10.18682/cdc.vi149.5515
4. Badarnah L (2012) Towards the living envelope: biomimetics for building envelope adaptation. Doctoral's Thesis, Technische Universiteit Delf, School of Technology and Architecture, Doctoral dissertation, Technische Universiteit Delf, Netherlands
5. Benyus J (2012) Biomimética: Inovação inspirada pela natureza, 6th edn. Pensamento Cultrix Ltda
6. Bimetal Facade System. Available online: https://www.adamheisserer.com/#/bimetal-facade/. Accessed 10 June 2022
7. Charpentier L, Cruz E, Nenov T, Guidoux K, Ware S (2022) Pho'liage: towards a kinetic biomimetic thermoregulating façade. In: Palombini F, Muthu S (eds) Bionics and sustainable design. Hong Kong, Springer, pp 367–401

8. Charpentier V, Hannequart P, Adriaenssens S, Baverel O, Viglino E, Eisenman S (2017) Kinematic amplification strategies in plants and engineering. Smart Mater Struct 26:1–30
9. DO|SU Studio Architecture. Available online: https://www.dosu-arch.com/bloom. Accessed 20 June 2022
10. Engineered Material Solutions (n.d.,) Thermostatic bimetal designer's guide. Wickeder Group, Attleboro, p 50
11. European Solar Shading Organization (2006) Energy saving and CO_2 reduction potential from solar shading systems and shutters in the EU-25, ESCORP-EU25, p 23
12. Fiorito F, Sauchelli M, Arroyo D, Pesenti M, Imperadori M, Masera G, Ranzi G (2016) Shape morphing solar shadings: a review. Renew Sustain Energy Rev 55:863–884. https://doi.org/10.1016/j.rser.2015.10.086
13. Foged I, Pasold A (2010) Performative responsive architecture powered by climate. In: Proceedings of the 30th annual conference of the Association for Computer Aided Design in Architecture (ACADIA), New York, USA, 21–24 October 2010
14. Fox M (2016) Interactive architecture: adaptive world, 1st edn. Princeton Architectural Press, p 176
15. Heisserer A, Nguyen K, Pinder T (2013) Embedded responsiveness. FALL. UTA ARCH 5670, Texas, p 12
16. InVert Self-Shading Window System: flipping energy efficiency on its head. Available online: https://www.tbm-designs.com. Accessed 22 Feb 2020
17. Jarjat P, Nenov T, Ware S (2020) Pho'liage ®: a biomimetic façade which increases building energy efficiency. Artbuild, Brussels, p 50
18. Kanthal AB (2008) Kanthal thermostatic bimetal handbook. Kanthal AB: Sweden, 2008, p. 126
19. Menges A (2012) HygroScope: Morfologia Meteorossensitiva. http://www.achimmenges.net/?p=5083
20. Oliveira M (2019) Towards a bio-shading system concept design methodology. Doctoral's thesis, Instituto Universitário de Lisboa, School of Technology and Architecture, Lisboa, Portugal, December 2019
21. Ritter A (2007) Smart materials: in architecture, interior architecture and design. Birkhäuser – Publishers for Architecture, Washington, USA, p 189
22. Sack-Nielsen T (2017) Performance through thickfolds: approaching climate-responsive behaviours through shape, materialisation and kinematics. Doctoral's thesis, Aarhus School of Architecture, Danish, 5th May 2017
23. Shapiro G (2020) Award: InVert self-shading windows flip energy efficiency on its head. Architect Magazine, 12th July 2020. https://www.architectmagazine.com/awards/r-d-awards/award-invert-self-shading-windows-flip-energy-efficiency-on-its-head_o
24. Sung D (2016) Smart geometries for smart materials: taming thermobimetals to behave. J Archit Educ 70:96–106. https://doi.org/10.1080/10464883.2016.1122479
25. Vazquez E, Randall C, Duarte J (2019) Shape-changing architectural skins a review on materials, design and fabrication strategies and performance analysis. J Facade Des Eng 7:91–102. https://doi.org/10.7480/jfde.2019.2.3877

Tarciana Andrade graduated in Design from UFPE in 2009, obtained her master's degree in Design from UFPE in 2014, focusing her research on design-driven innovation for the Creative Economy of Pernambuco. She completed her PhD in Design at FA-ULisboa in 2024, researching the development of bioinspired adaptive modules for façades that are self-actuated without electromechanical devices. She also served as the coordinator of the makerspace at Porto Digital and taught design and architecture courses in Pernambuco at UNIBRA and UNIFG.

José Beirão holds a degree in architecture (1989) from the Faculty of Architecture, TU Lisbon. He is the founder of Bquadrado architects (1998–2010), has a Master's degree in urban design (2005) from ISCTE-IUL, University Institute of Lisbon, and did his PhD in Urbanism and

Architectural engineering + Technology from TU Delft, Netherlands, in 2012. He is the coordinator of the Design and Computation Group at FA-ULisboa since 2016. His main research topics are CIM (city information modeling), spatial analysis, and urbanism and sustainability.

Amilton José Vieira de Arruda graduated in Industrial Design Product Project from UFPE and holds a Master in Design and Bionics from IED in Milan, a PhD in Ricerca in Disegno Industriale from the University of Milan Politecnico, a postdoctoral in Design and Bionics from IADE European University UNIDCOM Lisbon, and a postdoctoral in Bionics from the University Luigi Vanvitelli Naples—IT. He is currently associate professor IV at the UFPE Design Course and coordinates the BioDesign Lab.

Chapter 2
Utilization of Accessible 3D Technologies for the Study of Bionics in Jewelry Design

Giovanna Ribeiro de Oliveira, Mariana Kuhl Cidade, Ronaldo Martins Glufke, and Felipe Luis Palombini

Abstract Bionics, biomimetics, and biodesign are often associated with the employment of state-of-the-art technologies from the investigation to the application of natural characteristics in projects. In addition to a more sustainable approach, contemporary jewelry is focused on the disruptive application of unusual materials and processes, either by exploring new forms of inspiration or looking at ways to universalize its methods. In this sense, this chapter explores the use of accessible technologies and processes for the application of bioinspired design, intending to make it more reachable and affordable. Light microscopy, 3D scanning, and 3D printing are proposed as methods of observation and manufacturing of bioinspired designs, applied in the development of a contemporary jewelry collection.

Keywords Contemporary jewelry · Bionics · Digital manufacturing · Sustainability · Additive manufacturing · Photogrammetry · Affordable technologies · 3D technologies

---

G. R. de Oliveira
Industrial Design Undergraduate Course, Federal University of Santa Maria (UFSM), Santa Maria, RS, Brazil
e-mail: giovanna.ribeiro@acad.ufsm.br

M. K. Cidade (✉) · F. L. Palombini
NOVA Lab – Innovation and Sustainability in Design Lab – Department of Industrial Design, Graduate Program in Architecture, Urbanism, and Landscaping (PPGAUP), Federal University of Santa Maria (UFSM), Santa Maria, RS, Brazil
e-mail: mariana.cidade@ufsm.br; felipe.palombini@ufsm.br

R. M. Glufke
NOVA Lab – Innovation and Sustainability in Design Lab – Department of Industrial Design, Federal University of Santa Maria (UFSM), Santa Maria, RS, Brazil
e-mail: rglufke@ufsm.br

A. J. V. de Arruda, F. L. Palombini (eds.), *Biology, Biomimetics and Natural Design*, Environmental Footprints and Eco-design of Products and Processes,
https://doi.org/10.1007/978-3-031-65877-8_2

2.1 Introduction

Over the centuries, jewelry has developed a variety of different functions, which can carry symbolic, social, economic, and religious meanings, in addition to, naturally, having the function of adorning, beautifying, and also being considered a form of communication [1]. As a result, jewelry has been the subject of countless interpretations, being a multidisciplinary area that encompasses art, materials, techniques, processes, social issues, craftsmanship, and design [2, 3].

If, for a long time, the value of a jewel was synonymous with preciousness and nobility, today this idea is definitely being modified [4–6]. The materials and tangible preciousness of the pieces are no longer the only characterizing elements to define whether or not an object belongs to the sphere of luxury [7]. The material, regardless of whether it is gold, platinum, diamonds, emeralds, wood, polymers, ornamental rocks, natural fibers, ceramic materials, rubbers, or even residual materials retrieved from waste [8], is a choice of design and its creator, where the techniques, technologies, finishes, and concepts given to jewelry are developed with a focus on the changes that society presents, in terms of choices, skills, sustainable visions, social, cultural, and economic issues, among others [5]. This new mode of contemporary jewelry design, referring to creative processes, unusual material choices, multiple metal alloys, and technologies with a good finish, is used due to changes in different spheres that have occurred over the centuries.

New technologies are a turning point for the development of contemporary jewelry, such as the use of laser machines [9], water jets [10] additive manufacturing of metals [11, 12], and MSLA (Masked Stereolithography Apparatus) among others [13]. The so-called unusual materials have also been used, often, as the only material in the piece, as long as they present a high level of finishing, through the use of different technologies [14–16]. Furthermore, jewelry of this type is also being designed and manufactured, contrasting the presence of unconventional materials along with the addition of silver (Ag 950 or 925 alloys). This metal in Brazil, for example, is currently being widely used in industry due to its significant value, and the numerous forms of manipulation in the production process, such as the inclusion of textures and intentional differentiated oxidations (e.g., with the use of sulfur), also making it a contemporary material. In addition, copper-based alloys are also being included in the creative process of contemporary jewelry by exploring some of its unique visual characteristics, such as oxidation, leading to a beautiful patina [17, 18].

By challenging the prejudices of traditional jewelry and its association with wealth, social status, as well as cultural and religious positioning, contemporary jewelry uses new forms and techniques to emphasize the value of the piece [4, 19]. New sources of inspiration, creative processes, shapes and compositions, materials and means of artisanal and/or industrial manufacturing, allow unconventional methods to be extrapolated, which can bring a new difference to the pieces developed, giving greater freedom to those who produce and use them [8, 20]. New materials combined with diverse technologies enrich the meaning and value of a piece and can emphasize its sensorial experience for the user [5].

Bionics, Biomimetics, and BioDesign are often associated with the use of cutting-edge technologies from research to the application of natural characteristics in projects [1, 21, 22]. However, not everyone has access to advanced technologies when developing bioinspired projects, leading to the need to search for more feasible and accessible alternatives to investigate the natural world. In addition to a more sustainable approach, contemporary jewelry focuses on the disruptive application of unusual materials and processes, whether exploring new forms of inspiration or seeking ways to universalize its methods. Therefore, this chapter explores the use of two methods of contemporary jewelry production [20]: one considered more traditional and another more experimental. In the first one, an artisanal oxidation process on copper was realized, followed by wax machining and silver alloy investment casting. In the second one, new, yet accessible 3D technologies for 3D scanning and 3D printing were used. Those are proposed as methods for observing and manufacturing bio-inspired designs applied to contemporary jewelry. This chapter describes research in the NOVA Lab—Innovation and Sustainability in Design Lab, from the Federal University of Santa Maria (UFSM) under the research lines of Bionics, Contemporary Jewelry, Digital Manufacturing, and Sustainability.

2.2 Copper: Problematic Material in Brazil?

Copper (Cu) is a material used in several areas, thanks to its remarkable conductivity and malleability [23], being widely used since prehistoric times as an alloy in jewelry [24, 25]. For gold (Au) and silver (Ag) jewelry, copper is added to obtain better mechanical resistance, which is important for the durability of the pieces [26]. Due to its properties, copper ends up as one of the most used metals, being used, for example, in the manufacture of electrical wires and electronic components. However, as it is a raw material with increasing scarcity [27], it is important to think about its recyclability, as the metal can be reprocessed virtually countless times [28].

In Brazil, this material is becoming a problematic metal due to its inappropriate disposal and theft. Copper can be found in discarded electronics [29] and, abundantly, in public network wires, such as electrical, telephone, and signaling (Fig. 2.1). Many Brazilian cities have their wires exposed on the roads, causing disruption to the city's visibility in aesthetic terms, and often copper wire theft keeps them even more noticeable when partially and illegally removed. In Brazil, the equivalent of 4.72 million stolen electrical wires were recorded in 2022. For instance, in the city of Porto Alegre, in the state of Rio Grande do Sul, in the south of the country, around 11,400 m of copper cables were stolen across the city, from January to April 2023 [30], according to the registers of the Public Transport and Circulation Company (EPTC).

Fig. 2.1 Copper wire extracted from recovered electronics

Electronics at the end of their useful life are often discarded incorrectly in Brazil, a practice that has shown a considerable increase in recent years. In 2020, the sectoral agreement governed by Act No. 10.240 was published, which establishes standards for the implementation of a mandatory reverse logistics system for household electronic products and their components, with the objective of structuring, implementing, and operationalizing the system nationally [31]. The management entities responsible for operationalizing this system are the Electrical and Electronic Reverse Logistics Manager (Green Electron) and the Brazilian Association for Recycling of Electronic and Domestic Appliances (ABREE).

The majority of municipal solid waste (MSW) collected (61%) continues to be sent to treated landfills, with 46.4 million tons sent for environmentally appropriate disposal in 2022 [32]. On the other hand, areas of inadequate disposal, including landfills, are still in operation in all regions of the country, receiving 39% of the total waste collected, reaching a total of 29.7 million tons of inappropriate disposal [32]. According to information available in the first Annual Performance Report of the Reverse Logistics System (SLR) for electronics and their components for domestic use, there are currently 3417 reception points spread across 1224 municipalities in the 26 states of the Brazilian federation, plus the Federal District [32]. In 2021, around 1245 tons of electronic and household appliance waste were collected and disposed of in an environmentally friendly manner.

In design, copper is being added to multiple projects. In addition to being used as metal for the manufacture of various electronics, the aesthetic modification of its surface through a natural oxidation process is being explored. Copper has some of its visible characteristics as a reddish color, but when exposed to humid air it goes through a slow oxidation process, creating a surface layer of copper sulfate with a greenish hue, popularly known as patina [17, 33]. In this sense, this research sought to combine unusual materials, such as oxidized copper, with creative processes and different, more accessible technologies to develop a jewelry collection.

2.3 BiOxi Jewelry

This section presents the development of the BiOxi jewelry collection, which was aimed to combine sustainable issues, bio-inspired creative processes, and different technologies. In sustainable terms, copper extracted from electrical wires that were recovered from damaged electronics was used and recycled. As for different technologies, traditional investment casting of a silver alloy over a CNC machined wax as well as experiments with additive technology, or 3D printing, were followed.

For the electrical wires, a manual stripping process was followed to remove the polymeric coating. Once stripped, the copper wires were recycled following a semi-industrial process. First, the material was melted with an oxyacetylene torch, shaped into an ingot, and rolled into a sheet format. Afterwards, an artisanal oxidation process was carried out, using accessible solutions containing combinations of water (60 ml), salt (36 g), and vinegar (60 ml). The recycled copper sheets, after being soaked for 30 min, were tested with different compositions of water and salt to check for the acceleration of the oxidation. In addition, a solution of vinegar and salt was also tested, with a soaking time up to 1 h 20 min. As a result, specific oxidation occurred scattered in some parts of the copper sheets. Salt and vinegar solution was chosen due to a more uniform oxidation throughout the piece, forming a patina that covered the surface. Part of the recycling process by artisanal casting and oxidation is shown in Fig. 2.2.

Once the oxidized copper sheets were selected as the main featured material for the proposed piece, the project was followed by means of a methodology of bionics and contemporary jewelry design [20, 34]. Bionics was added to this jewelry collection as a way to find natural sources of inspiration for obtaining technical design solutions, regarding aesthetic, formal, structural, and functional aspects [35]. Moreover, the project development process was associated with the use of technologies, which were selected following the fundamentals of being feasible and accessible. Therefore, the main purpose of this project was to establish practical and viable ways to include a Biomimetic approach in a different application, such as a jewelry collection, showing that the utilization of state-of-the-art technologies is not mandatory for taking inspiration from nature in a creative process.

The creative process was inspired by textures, such as the result of copper oxidation, and movements that nature provides us. In addition to the texture of the copper oxidation itself, elements of nature were explored, with the verification and visualization of shapes and movements. For this purpose, mushrooms of the *Armillaria tabescens* (Scop.) Emel species (Physalacriaceae) were selected. This species was selected due to its colors, shades, shape patterns, and particular interior structure, as shown in Fig. 2.3. Mushrooms from this family usually have a distinct texture on their top, similar to that of the patina obtained from the oxidation of copper sheets. Although there are many species of mushrooms, they usually have a typical morphology, consisting of a cap (pileus), a stem (stipe), and gills (lamellae) on the underside of the cap. The pileus is the upper part of the mushroom, which contains the fertile area, that is, the structure where spores are produced, which is called the

Fig. 2.2 Copper casting and oxidation process

Fig. 2.3 *A. tabescens* (Scop.) Emel. Mushrooms

hymenium. This structure is normally facing downwards, facilitating the release and dispersion of the spores when they are mature. The stipe is normally partially buried in the ground and its structure supports the entire hat.

In order to visualize these aesthetic, structural, and functional aspects, some technologies were used in the development of the collection, being divided into technologies for observation, such as light microscopy and 3D scanning using a cellphone application, in addition to CNC production technologies such as subtractive and additive manufacturing. Ever since the observation of cork by pioneer studies of Rober Hooke, light microscopy represents the foundation technology for the observation of fine details in biological samples [36]. On the other hand, despite being considered substantially more advanced, 3D scanning has become an ever-increasing accessible method for a number of projects, when obtaining volumetric meshes from real objects is required. This is due to the significant improvements in the cameras of modern cell phones, their processing capabilities, and the development of free 3D scanning app alternatives.

Regarding CNC production technologies, subtraction manufacturing is a set of some of the most important tools in product design, mainly those related to machining, such as turning, boring, drilling, milling, and sawing, as well as those related to laser, primarily cutting and engraving. As for additive manufacturing technologies, commonly known as 3D printing, despite being relatively new in the realm of production techniques, such processes are becoming more affordable and utterly present in all stages of product development. Especially, fused filament fabrication techniques are regularly used for prototyping and mock-up manufacturing due to the number of commercially available alternatives as well as the low cost of both the equipment and its inputs, such as ABS or PLA polymeric filaments.

These were some of the proposed methods for the manufacturing of this bioinspired contemporary jewelry collection. Along with these technologies, physical paper mock-ups were also employed to help in understanding the general shapes and dimensions of the selected bioinspired geometries. This process was followed to allow a physical visualization of the forms since it is a low-cost method for producing 3D shapes. After that, the defined geometries were parameterized, and the project was modeled in 3D by means of CAD software [34].

Firstly, some samples of mushrooms were collected in the city of Santa Maria (Rio Grande do Sul, Brazil), and these were observed under light microscopy. In addition to seeking a better understanding of the biological characteristics, these photomicrograph observations stimulate creativity in projects, offering images with rich and often unexpected details about the object in question. In order to perform the light microscopy acquisitions, XTS–20 LBX (Nova® Optical Systems, Piracicaba, Brazil) stereoscope microscopy with zoom as well as CX21 (Olympus® Co., Tokyo, Japan) transmitted light microscope with multiple objectives were used. Both microscopy equipment are located in the Structural Plant Laboratory in the Federal University of Santa Maria (LABOTE/UFSM). Observations were used in this collection to assist the creative process of the pieces, as visualized in Fig. 2.4.

The process began with the registering of images from the microscopy visualization of the mushroom, followed by the creative generation of ideas by means of a preliminary sketching. These were then refined to a desired shape. It is noteworthy that, given the fact that a jewelry piece requires particular care to the usability aspects of the final model, during the creative processes, ever since the first sketches

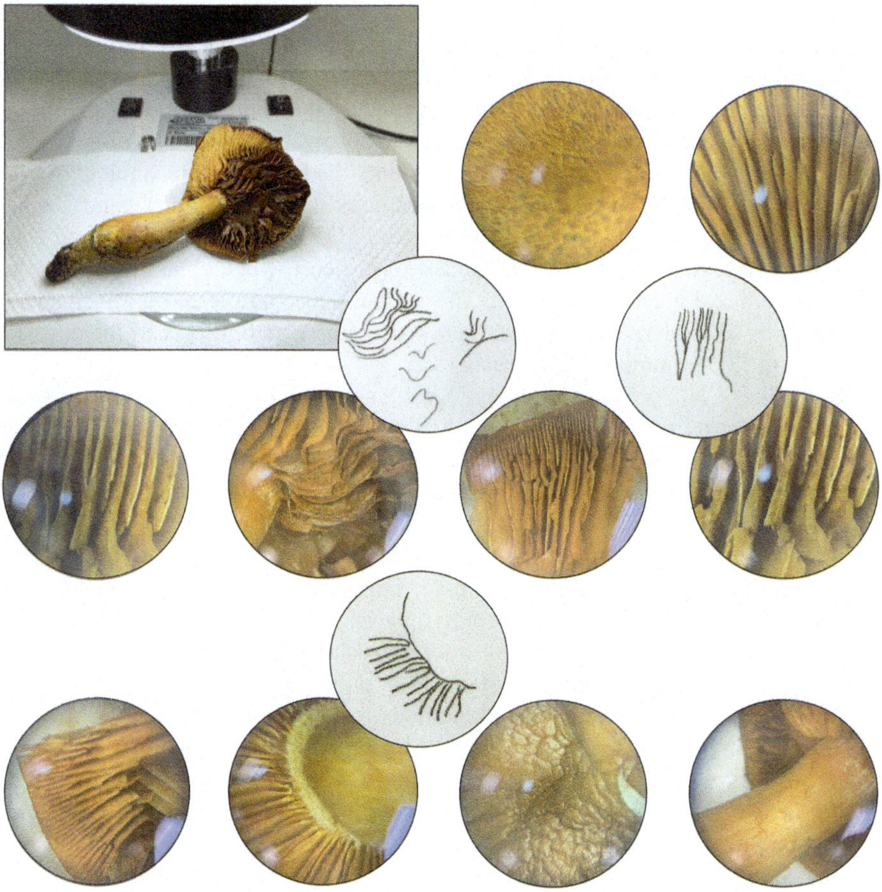

Fig. 2.4 Photomicrographs of the mushroom and preliminary sketches

were drawn, the whole process considered overall features such as dimensions and weight, besides possible ergonomics issues.

With the sketches developed, it was selected that the first piece designed for the proposed jewelry collection would be an earring. Therefore, Rhinoceros® 3D (McNeel®, Seattle, USA) CAD software was used for a more precise 3D modeling of the jewelry. In the software, in addition to exact dimensions, the main components and materials, such as silver and copper, can be modeled and set to either visually preview how the earrings would look, as well as to estimate some of the ergonomic issues prior to manufacturing. In Fig. 2.5, the rendering of the pair of earrings is shown.

After 3D modeling, the earrings went through the subtractive manufacturing process, using CNC machining in wax. Ferris® (Freeman® Manufacturing & Supply Co., Avon, OH, USA) wax and the 4-axes Modela MDX-40A (Roland® DG Co., Hamamatsu, Japan) machine tool was used in the machining process aimed at

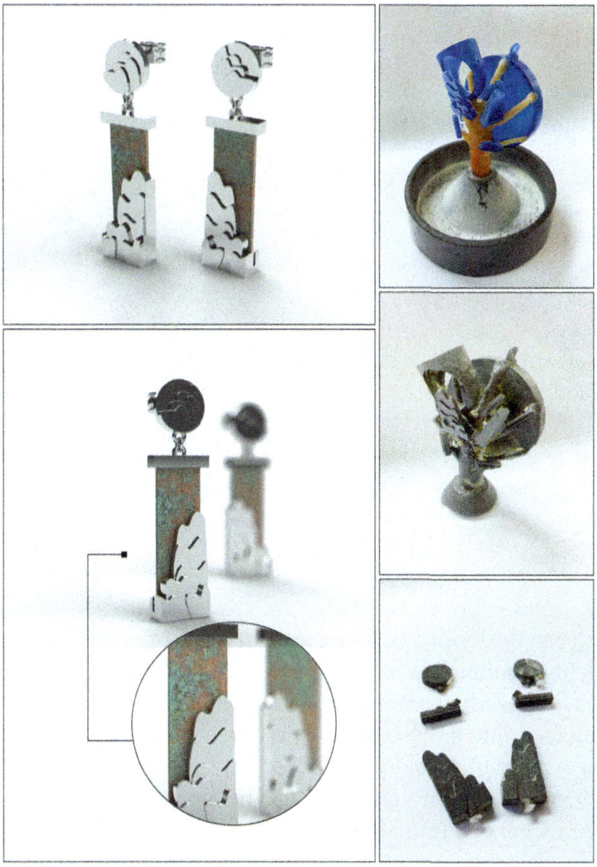

Fig. 2.5 Jewelry manufacturing process: from 3D modeling to casting and final pieces

investment casting in metal. In this process, only the parts of the earring that would be made of silver (Ag 950) were machined. The copper used, as previously mentioned, was manufactured into sheet metal form and subsequently oxidized for a distinct patina finish. Copper wires retrieved from e-waste were also used in the formation of the silver alloy, which contains 5% of copper in weight. After carrying out the silver casting process, the metal parts went through an artisanal finishing process, with the subsequent steps of sanding, polishing, and assembly. In Fig. 2.6, the finished jewelry piece is shown.

After the more traditional physical manufacturing of the earring from the BiOxi collection, two new technologies were used for a second, more experimental piece: 3D scanning and additive manufacturing. The main objective was to show how simple, yet unconventional 3D technologies can be used to combine a bioinspired approach to contemporary jewelry design.

Fig. 2.6 BiOxi earrings

3D scanning can be divided between several types of technologies for data acquisition, and photogrammetry is becoming one of the most accessible methods. The technology lies in the passive capturing of visible light via photos at different angles that are later merged into a 3D volume by means of some algorithm. There are several proprietary photogrammetry-dedicated equipment, known as 3D scanners, which can vary in cost and thus generally can be unreachable for some. On the other hand, due to the recent evolution of the overall quality of cellphone cameras, several apps with a free-use approach have been developed, making photogrammetry-based 3D scanning more accessible, such as KIRI Engine (Fig. 2.7).

For this research, the collected mushroom was first dried in order to better highlight some natural curves of the gills below the pileus of the sample. This dedicated KIRI Engine app was selected due to the fairly good quality of its final 3D scanned meshes, as well as due to its free-of-charge availability in the basic plan, despite also having a subscription-based Pro option. The free, basic module was used, which allows users to take up to 70 photos of a subject, send them to the server where they are processed, and the app returns a 3D model free to download (up to 3 free exports per week). The app generated images from all the surface areas of the sample and was exported as an OBJ file. Such file extension not only encompasses the coordinates of the generated mesh but also includes image data retrieved while scanning, thus contributing to a better understanding of the details of the sample's morphology. After that, the file was imported into Rhinoceros®, as visualized in Fig. 2.7.

After loading the OBJ file into Rhinoceros®, some of the geometries of interest were parametrized [34], and the process is shown in Fig. 2.8. The mushroom gills were the part that stood out most in the scan. As a result, these curvilinear shapes,

Fig. 2.7 3D scanning of a mushroom sample via KIRI Engine and the resulting OBJ mesh in Rhinoceros® 3D

as if they were paths, were simplified in Rhinoceros® and captured through curves, being extruded as surfaces. These simplifications with curves and surfaces lead to the generation of two main modules from two distinct regions of the mushroom. Afterward, for a greater understanding of these traced geometries, physical mock-ups with paper were developed. The parametrized patterns of the curves, representing the regions of the gills, were marked with a laser equipment. EXLAS-X4 (Jinan XYZ® Machinery LLC, Jinan, China) CO_2 laser equipment with a max output power of 100 W with a speed of 10 mm/s was used to engrave the surface of medium density fiberboards (MDF) plates, replicating the curves designed in the software.

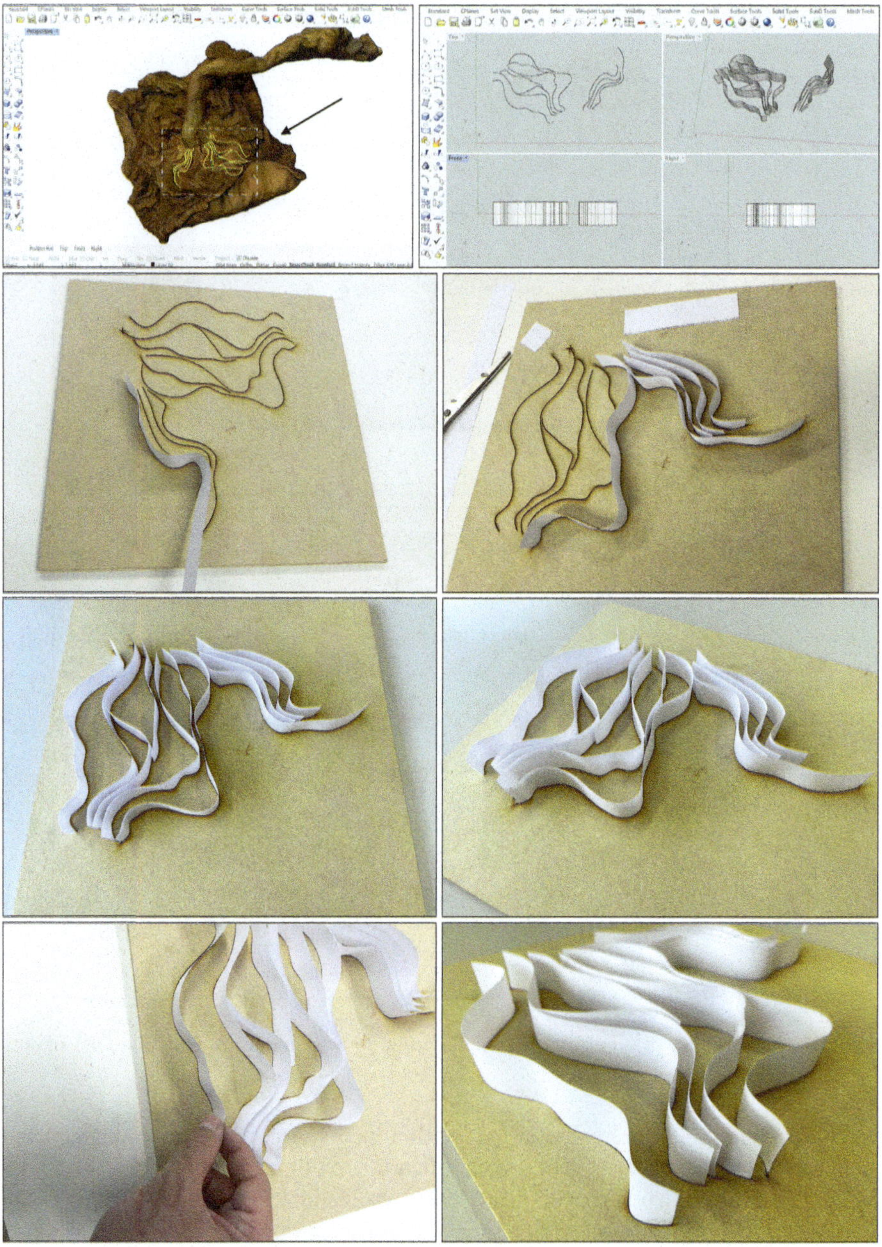

Fig. 2.8 Simplification and parametrization of forms, generation of modules, and construction of physical mockups

Through these mock-ups, the macroscopic modules derived from the movement of the curves taken from the anatomy of the mushroom gills that were captured with 3D scanning were sought, so they could be transmitted and adapted in a volumetric manner (Fig. 2.8). Module repetition studies could be developed and visualized both in physical mock-ups as well as virtually, in the 3D software. The repetition in circular and elliptical formats was defined while considering common jewelry application formats, such as pendants. One of the current tool possibilities that the newer versions of Rhinoceros® 3D presents is SubD modeling. This tool can be explored to generate new types of organic geometries by means of editable and highly accurate shapes. This was then also used to explore new formats in some parts that make up the modules.

Finally, in order to exemplify in this jewelry collection a different type of affordable technology, an entry-level Additive Manufacturing process was used instead of a conventional jewelry casting manufacturing procedure. Therefore, the modeling carried out by SubD in Rhinoceros® was developed and applied to a pendant with geometric modules in a circular format, for the use of 3D printing technology (Fig. 2.9).

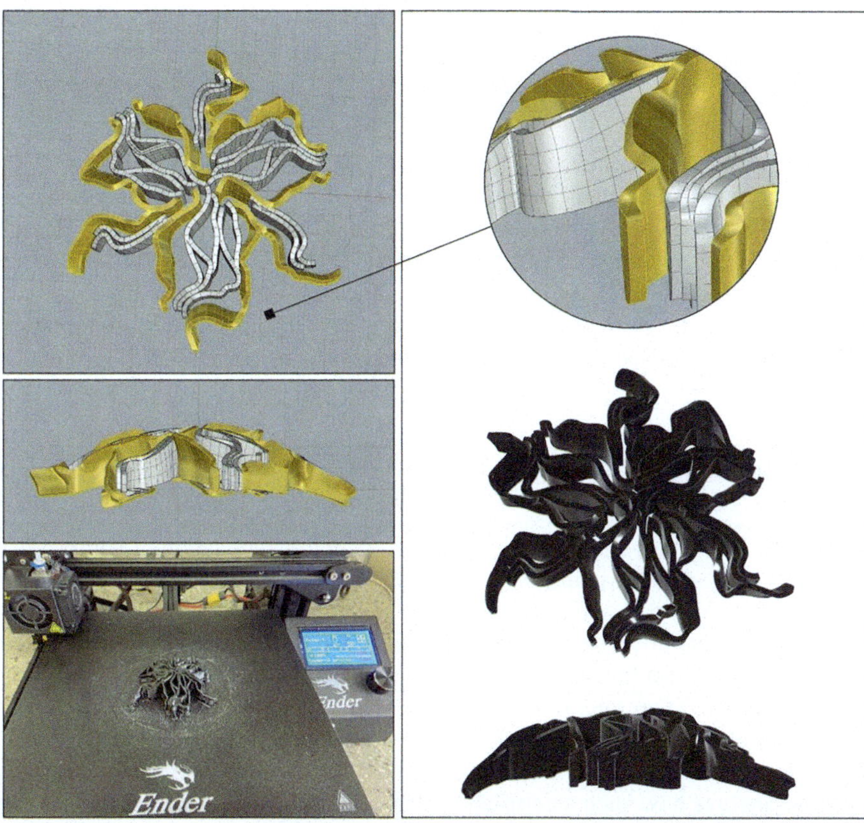

Fig. 2.9 Software 3D modeling and 3D printing of the pendant

Fig. 2.10 BiOxi pendant

The 3D modeled geometry of the pendant was then exported into an STL file which was then sliced using Cura (Ultimaker® B.V, Utrecht, The Netherlands), using a 0.12 mm layer height, in order to be printed with a PLA filament. Ender 3 (Creality® 3D Technology Co., Ltd., Shenzhen, China) 3D printer was used, being an accessible, entry-level equipment that balances reasonably good print quality with a low cost. The second manufactured piece from the BiOxi collection is shown in Fig. 2.10.

2.4 Conclusions

Bionics, Biomimetics, and BioDesign are based on the application of natural characteristics in design activities, which may include attributes aimed at aesthetics, function, and structure, among others. Since its first conceptual idealizations, it has sought to integrate different research fields and areas of knowledge to obtain essential information about the characteristics of natural elements and how to use them in design solutions. Therefore, it is essential that the development of bioinspired projects can be carried out with multidisciplinary teams, with skills ranging from biological areas (to understand the source element of inspiration), through technical areas (to analyze how such an attribute produces the desired effect), up to the creative and technological areas (to apply it effectively in a project).

This chapter explored the application of accessible technologies aimed at the combination of bioinspired design and contemporary jewelry. Both projects were creatively based on the geometries of a mushroom specimen, which was used to derive general information regarding the desired shapes. The first one utilized light microscopies taken from the sample, utilizing 3D modeling and traditional jewelry manufacturing techniques, such as silver alloy investment casting on a machined wax. The second one explored the use of 3D scanning via an app-based photogrammetry solution, whose 3D geometry was exported and studied with the construction of a physical model. Finally, the chosen pattern was modeled in 3D and printed using an entry-level additive manufacturing equipment.

Despite having great potential for integrating areas, projects based on bionics are often associated with state-of-the-art technologies and, often, unaffordable observation and analysis techniques, which would end up hindering access to one of the main steps in the development of a bioinspired project. It is clear that cutting-edge, sophisticated equipment and techniques are always stimulating to include when investigating some natural element, however, this cannot be seen as a limiting and restrictive factor in research. In this sense, it is interesting to present how new, yet accessible, observation and manufacturing processes can be explored to be included in the production of a contemporary jewelry collection.

References

1. Cidade MK, Perini JT, Palombini FL (2022) Bionics for inspiration: a new look at Brazilian natural materials for application in sustainable jewelry. In: Palombini FL, Muthu SS (eds) Bionics and sustainable design. Springer, Singapore, pp 195–223
2. Untracht O (2011) Jewelry concepts & technology. Doubleday, New York
3. Miller J (2016) Jewel: a celebration of earth's treasures. Dorling Kindersley Ltd., London
4. Scarpitti C (2021) The contemporary jewelry perspective. Meanings and evolutions of a necessary practice. J Jewellery Res 4:59–76
5. Cidade MK, Palombini FL, Palhano AP, Melchiors A (2021) Experimental study for the valorization of polymeric coffee capsules waste by mechanical recycling and application on contemporary jewelry design. In: Muthu SS (ed) Sustainable packaging. Springer, Singapore, pp 85–110
6. Ba'ai NM, Hashim HZ (2015) Waste to wealth: the innovation of areca catechu as a biomaterial in esthetics seed-based jewelry. In: Proceedings of the International Symposium on Research of Arts, Design and Humanities (ISRADH 2014). Springer, Singapore, pp 373–381
7. Cappellieri A, Tenuta L, Testa S (2020) Jewellery between product and experience: luxury in the twenty-first century. In: Gardetti MÁ, Coste-Manière I (eds) Sustainable luxury and craftsmanship. Springer, Singapore, pp 1–23
8. Palombini FL, Cidade MK (2022) Lixo Invisível: Contribuição do Design para Recuperação de Materiais Problemáticos. MIX Sustentável 9:17–26. https://doi.org/10.29183/2447-3073.MIX2023.v9.n1.17-26
9. Cidade MK, Palombini FL, Duarte L d C, Paciornik S (2018) Investigation of the thermal microstructural effects of CO 2 laser engraving on agate via X-ray microtomography. Opt Laser Technol 104:56–64. https://doi.org/10.1016/j.optlastec.2018.02.002

10. de Abreu e Lima CE, Lebrón R, de Souza AJ et al (2016) Study of influence of traverse speed and abrasive mass flowrate in abrasive water jet machining of gemstones. Int J Adv Manuf Technol 83:77–87. https://doi.org/10.1007/s00170-015-7529-9
11. Milewski JO (2017) Additive manufacturing metal, the art of the possible. In: Springer series in materials science. Springer, Cham, pp 7–33
12. Abisuga OA, de Beer D (2023) The opportunities of additive manufacturing in the creative industries: a bibliometric analysis. Progress in Additive Manufacturing 8:1179. https://doi.org/10.1007/S40964-022-00392-6/TABLES/3
13. Yap YL, Yeong WY (2014) Additive manufacture of fashion and jewellery products: a mini review. Virtual Phys Prototyp 9:195–201. https://doi.org/10.1080/17452759.2014.938993
14. SDN de S M, Cidade MK (2019) SUSTENTABILIDADE E JOALHERIA: RECICLAGEM DE EPS PARA APLICAÇÃO EM JOIAS. MIX Sustentável 5:27–39. https://doi.org/10.29183/2447-3073.MIX2019.v5.n4.27-39
15. Guerra AL, Cidade MK (2019) Design e Joalheria: desenvolvimento de uma joia com óleos essenciais para o alívio de sintomas alérgicos. Design e Tecnologia 9:115–130. https://doi.org/10.23972/det2019iss18pp115-130
16. Cidade MK, Palombini FL (2023) Experimental projects with problematic materials in jewelry education. In: ENSUS2023 – XI Encontro de Sustentabilidade em Projeto. Grupo de Pesquisa Virtuhab/UFSC, pp 362–374
17. Leygraf C, Chang T, Herting G, Odnevall Wallinder I (2019) The origin and evolution of copper patina colour. Corros Sci 157:337–346. https://doi.org/10.1016/J.CORSCI.2019.05.025
18. Corti CW (2016) Special colors of precious metal jewelry: present and future. In: The Santa Fe Symposium on Jewelry Manufacturing Technology 2016. Santa Fe, pp 165–198
19. Skinner D (2013) Contemporary jewelry in perspective. Lark Books, New York
20. Cidade MK, Palombini FL (2022) Design de joias: proposição de metodologia para ensino voltado ao mercado joalheiro. Design e Tecnologia 12:57–72. https://doi.org/10.23972/det2022iss24pp57-72
21. Palombini FL, Muthu SS (2022) Bionics and sustainable design. Springer, Singapore
22. Palombini FL, Nogueira FM, de Oliveira BF, de Araujo Mariath JE (2022) Two-way bionics: how technological advances for bioinspired designs contribute to the study of plant anatomy and morphology. In: Palombini FL, Muthu SS (eds) Bionics and sustainable design. Springer, Singapore, pp 17–44
23. Ashby MF (2011) Materials selection in mechanical design, 4th edn. Butterworth-Heinemann, Burlington
24. Hesse RW (2007) Jewelrymaking through history: an encyclopedia. Greenwood Press, Westport
25. Raub CJ (1995) The metallurgy of gold and silver in prehistoric times. In: Prehistoric Gold in Europe, pp 243–259. https://doi.org/10.1007/978-94-015-1292-3_16
26. Kliauga AM, Ferrante M (2009) Metalurgia básica para ourives e designers: do metal à jóia. Blücher, São Paulo
27. Wang C, Huang X, Lim MK et al (2020) Mapping the structural evolution in the global scrap copper trade network. J Clean Prod 275:122934. https://doi.org/10.1016/J.JCLEPRO.2020.122934
28. Gómez F, Guzmán JI, Tilton JE (2007) Copper recycling and scrap availability. Resour Policy 32:183–190. https://doi.org/10.1016/j.resourpol.2007.08.002
29. Cândido LHA, Kindlein Júnior W, Palombini FL (2015) Indicators for the feasibility of WEEE processing plants in Brazil: a relationship between virgin mineral copper ore and the recycling of consumer electronic products. In: Design e Tecnologia, pp 59–65
30. Trezzi H, Abati L, Grizotti G, Oliveira G (2023) O que está por trás do furto de cabos em Porto Alegre; quantidade de fios levados triplicou em dois anos. In: GZH. https://gauchazh.clicrbs.com.br/grupo-de-investigacao/noticia/2023/04/o-que-esta-por-tras-do-furto-de-cabos-em-porto-alegre-quantidade-de-fios-levados-triplicou-em-dois-anos-clgqx21hl007j0177vwqdteg3.html. Accessed 21 Sept 2023

31. MMA – Ministério do Meio Ambiente – Brasil (2022) Plano Nacional de Resíduos Sólidos. Brasília, DF
32. ABRELPE (2022) Panorama dos Resíduos Sólidos no Brasil 2022. ABRELPE, São Paulo
33. Ashby MF, Kara J (2013) Materials and design: the art and science of material selection in product design. Elsevier Science & Technology
34. Palombini FL, Cidade MK, Magris DA, Ghedini JVS (2021) Práticas projetuais transdisciplinares entre design e biologia: metodologia prática para o ensino de biônica. Revista Educação Gráfica 25:245–257
35. Kindlein Júnior W, Guanabara AS (2005) Methodology for product design based on the study of bionics. Mater Des 26:149–155. https://doi.org/10.1016/j.matdes.2004.05.009
36. Palombini FL, Cidade MK, de Oliveira BF, Mariath JE d A (2021) From light microscopy to X-ray microtomography: observation technologies in transdisciplinary approaches for bionic design and botany. Cuadernos del Centro de Estudios en Diseño y Comunicación 149:61–74. https://doi.org/10.18682/cdc.vi149.5516

Giovanna Ribeiro de Oliveira is an undergraduate student in Industrial Design in the Federal University of Santa Maria (UFSM). She is a volunteer researcher in the Innovation and Sustainability in Design Lab (NOVA Lab) at UFSM.

Mariana Kuhl Cidade is an assistant professor in the Department of Industrial Design (DDI) and in the Graduate Program in Architecture, Urbanism, and Landscaping (PPGAUP) at the Federal University of Santa Maria (UFSM). Dr. Cidade graduated in Industrial Design and holds an M.Sc. and PhD in Design (PGDesign) from the Federal University of Rio Grande do Sul (UFRGS). Her research includes sustainability, contemporary jewelry, materials recycling, bionics, and product design. She is associated with the Innovation and Sustainability in Design Lab (NOVA Lab) at UFSM.

Ronaldo Martins Glufke is an associate professor in the Department of Industrial Design (DDI) at the Federal University of Santa Maria (UFSM). Dr. Glufke graduated in Industrial Design and holds an M.Sc. in Design from the Federal University of Santa Catarina (UFSC) and a PhD in Design from the Università degli Studi di Firenze (UNIFI). His research includes Product Design, Design Management, Design and Training, Surface Design and Ceramic Coatings. He is associated with the Innovation and Sustainability in Design Lab (NOVA Lab) at UFSM.

Felipe Luis Palombini is an assistant professor in the Department of Industrial Design (DDI) and in the Graduate Program in Architecture, Urbanism, and Landscaping (PPGAUP) at the Federal University of Santa Maria (UFSM). Dr. Palombini holds an M.Sc. and a PhD in Design (PGDesign) from the Federal University of Rio Grande do Sul (UFRGS). His main research topics are focused on bionics, sustainability, digital manufacturing, and materials characterization. He is associated with the Innovation and Sustainability in Design Lab (NOVA Lab) at UFSM, the Design and Computer Simulation Research Group (DSC), and the Plant Anatomy Laboratory (LAVeg), at UFRGS.

Chapter 3
Ecological Materialities: Exploring the Entanglement of Nature, Design, and Technology in Contemporary Materials

Enza Migliore and Matteo Convertino

Abstract The essay proposes an analytical and operative framework for designing future materials. It discusses emerging materiality's ontological and methodological identity through the stance of recently established post-anthropocentric theories and practices and the authors' design explorations. It shows the experiments, prototypes, and reflections arising from ongoing research projects to discuss the crucial role of design in establishing practices and perspectives on the materials of our surroundings informed by natural behaviors and enhanced by digital technology. This study critically reviews and juxtaposes different aspects of the authors' projects, *Breath*, *A Grain of*, and *Future Archaeologies*, to summarize the reflections and directions. The projects investigate the concurrence of nature and design in creating and altering the built environment and propose digital technologies as the medium to design new opportunities for it. We adopt cross-disciplinary methods and speculative practices of research through design, scientific sampling, microscopic observation, and archaeology-inspired processes. This essay derives directions and opportunities for designing material systems based on the entanglement of natural and artificial, human and nonhuman, and informed by the implicit logic of ecology.

Keywords New materialism · Digital ecologies · Other-than-human · Conviviality · Ecological design · Ecological materialities

---

E. Migliore (✉)
SUSTech School of Design, Shenzhen, China
e-mail: emigliore@sustech.edu.cn

M. Convertino
fuTuRE EcoSystems lab (TREES), State Key Lab for Ecological Remediation and Carbon Sequestration, Tsinghua Shenzhen International Graduate School, Tsinghua University, Shenzhen, China

A. J. V. de Arruda, F. L. Palombini (eds.), *Biology, Biomimetics and Natural Design*, Environmental Footprints and Eco-design of Products and Processes,
https://doi.org/10.1007/978-3-031-65877-8_3

3.1 Introduction

In nature, everything makes sense. In the built environment, we need to make sense of things. This evidence calls for profound reflections and ontological transformations in a planetarian crisis driven by human activities.

Design is evolving from an anthropocentric to a bio-inclusive discipline to align with grander life's purpose and motivation. This includes transcending the sustainability narrative to embrace the more complex ecology plot. Ecology is a systemic framework that considers the entanglement of living and nonliving beings and the environments, explored through a temporal and situated perspective. Ecology is based on relational factors to maintain balance [1, 2]. In *Towards an Ecology of the Mind*, Gregory Bateson [3] writes that ecology is a way of looking at the relationships between the "messages and pathways outside the body."

Materials signify a central issue in this emerging design and philosophical framework since they are the tangible interfaces of every designed interaction, the infrastructures for any use, function, and application, and the codification of natural resources into artificial means. Through projects and experiments focused on materiality, this chapters corroborates the argument that the boundaries of design practice, education, and research are expanding beyond the dichotomies of nature-culture, human-nonhuman, indoor-outdoor, digital-physical, and a singular discipline [4, 5].

The authors propose an ontological and methodological identity for future material systems design based on ecological and bio-centered approaches. It is ecology-informed, technology-enabled, cross-disciplinary, and methodologically undisciplined.

In this chapter, biodesign is the cross-disciplinary reference of contemporary practices for codesigning with nature; new materialism is the philosophical perspective on human-nonhuman entanglement; digital ecologies is the analytical scheme for exploring ontologically and ethically the digital entanglement in nature; ecological design is suggested as the operative framework; and ecological materialities is the ontological and methodological identity of future material design we propose based on the integration of the previous references.

3.2 Backgroud

3.2.1 Biodesign, the Key to Ecology

Biodesign [6, 7] is becoming the new paradigm where designers adopt living organisms, such as bacteria, fungi, and plants, that transfer ecological functions and information into devices, materials, and systems to achieve a positive footprint and regenerative procedures. Biodesign's agencies are based on a post-human and post-anthropocentric perspective, and its narrative activates social and cultural consciousness. It proposes new ways of thinking about materials, objects, and manufacturing processes, and new frameworks for interdisciplinary—and even interspecies—collaborations. This shift is science-informed and technology-enabled, contributing to the advancement of the theories and perspectives of New Materialism [4].

As Paola Antonelli outlines in the foreword of the book *Grow the Future: Visions of Biodesign*' [8], this emerging discipline may be mistaken for another, even more explicit, human agency of power over nature since it adopts living organisms as design materials instead of wood and metal. However, Biodesign activates a new way of looking at nature and other beings, getting closer to them, knowing, living together, and acknowledging their ecological role. Bees, plants, and microbes are the new co-agents of the design process, and humans must value their experience and intelligence. Beyond any breakthrough in design by biotechnology, Biodesign's main goal is to restore connections, empathy, and appreciation between humans and nonhumans because the more evident impact of human mastery over nature—and, today, biology—is precarity, crisis, and extinction. Humanity can achieve the most superior, sophisticated, and intelligent innovation. Still, without respect, friendship, and gratitude for the otherness, it can turn against itself. The challenge of Biodesign calls for interdisciplinary and interspecies encounters to become the paradigm of our future. The education of the next generation of designers must be imbued with this approach and purpose. *Biodesign Challenge*, founded by Daniel Grushkin, is an education program for high school and college students that introduces the intersection of biotechnology, art, and design and connects students to a global network of leaders in academia, industry, and entrepreneurship, defining the field of Biodesign.

Through data-driven experiments and technologies, speculative and critical perspectives, students tackle human awareness and perception of their being as opposed to or disconnected from nature and propose projects where animals, plants, and unicellular organisms function and survive symbiotically, and their interdependence generate reflections and visions for new biopolitics models.

3.2.2 New Materialism: Beyond Binary Taxonomies

To navigate and comprehend the complexity of life and society, design research and practice must go over the binary material categorization based on generalized and simplified taxonomies and market demand.

The design of our daily life spaces and devices strongly depends on needs, activities, and habits that arose decades ago, after cultures and practices based on capitalism and colonialism [9]. Human beings confined animals and plants in rigidly designed spaces of control and surveillance; engineers and architects used inert, artificial, homogenous materials to efficiently protect our houses from the outdoors; designers ideated mass-produced objects, wearables, services, and experiences that depend on heavily polluting, or rare resources and materials.

These design decisions made on species and ethnic superiority, enabled by technical frigidity and driven by market monopolies, are now dismantled by a planetarian crisis.

The philosophical approaches of new materialism offer novel standpoints to address materialities as inclusive of "[...] human bodies; other animate organisms; material things; spaces, places and the natural and built environment that these contain; and material forces including gravity and time. Also included may be abstract concepts, human constructs, and human epiphenomena such as imagination, memory, and thoughts; though not themselves 'material,' such elements can produce material effects" [4].

The new materialism approach contributes to an ontological, epistemological, and methodological convergence between social and natural science [10]. It challenges several dichotomies central to the humanities and social sciences, such as the distinction between the natural and social worlds, human and nonhuman entities, animate and inanimate beings, and the mind and matter [11, 12]. By questioning the conventional distinction between subjectivity and objectivity, it faces a reevaluation of data collection, analysis, and reporting methods [13].

The materiality addressed in these new materialisms is plural, open, complex, uneven, and contingent [14]. In Biodesign, materials are empowered with vitality or liveliness instead of inert and passive matter. [15].

New materialism explores how things other than humans, such as tools, technology, or buildings, can be social "agents," emancipating the affective capacities of the nonhuman [16].

3.2.3 Digital Ecologies

Digital technologies facilitate human-nonhuman interactions, from sensors and cameras for wildlife observation to machines used for microscopic analysis, computer-aided design, scanning, and advanced manufacturing technologies that can replicate organic forms that were once considered impossible. Environments are constantly sensed, recorded, represented, and commodified via digital technologies [17, 18].

Digital entanglement is a condition of the contemporary socio-ecological era. Turnbull et al. [17] propose "digital ecologies"' as an analytical framework to help understand the ethical, sociological, and environmental impact of the technological present [19] and "[...] to give researchers the toolkit with which they can understand the nuances and multiplicity of digital entanglement in specific empirical contexts." [17, 18]. The goal is to advance an ethical framework for living together in digitally mediated, more-than-human worlds.

Digital ecologies frame the integration of humans within "the environment." This explorative and critical structure helps transcend the binary between the eco-skepticism movements arguing for technology as a threat to human-nature connection and public ecological disengagement [20–22] driving humanity's "extinction of experience" [23] and "digital solutionism" [24] embracing the optimistic idea that digital technologies can provide new forms of nature reconciliation.

Digital ecologies clarify the nature and consequences of digital entanglement to guide future research in this emerging field across three main factors: Materialities, Encounter, and Governance [17].

Turnbull et al. [17], referring to ecological sensors, drones, cage cameras, digital microscopes, etc., unveil the heavy material side of technologies hidden under the cloud metaphor and highlight their biopolitical effects when applied to environmentalism and ecological restoration. However, they not only present a critique of unpredicted and commodified consequences of digital entanglement but also enlighten a series of unprecedented opportunities for human-nonhuman conviviality, offer valuable case studies to inspire progressive ethical directions and practices, and stress the focal importance of determining responsibility and obligations of ecological human activities. For instance, they discuss how digitization is responsible for environmental issues related to material processing of extraction, manufacturing, and disposal and all the associated social and political consequences. Digital materialities extend beyond metals and minerals, too, contributing to the production of plastic, radioactive waste, and other forms of pollution [25, 26]. They argue that emphasizing the discourse around the materiality of digital technologies can help the research visualize, identify, and explore new types of spatialities, materialities, relations, and politics. The discussion around digital encounters and governance elucidates the complexity of the related progress, consequences, and conditions. Digitization enables new intra-species encounters, favoring a mediated observation and emotional intimacy and informing advanced policies to restore biodiversity and environmental remediation. Screens, in this sense, are defined as ecological since they forge connections within bodies and environments [27].

Conversely, digital encounters may tend to simplify and generalize biological processes and behaviors due to their virtual decontextualization, and the limitation in sensual experience may weaken the bond between humans and nonhumans. Moreover, the digital interface embeds the "observer and observed" role, thus defining a gamification of ecological relationships. Consequently, most of the research and practice in this field still favors the user experience against the conviviality among species and agents. Surveillance technologies and live stream data collection are guiding policymaking for environmental purposes. Nevertheless, data are becoming the new commodification of ecological values and, if sold to third parties, can determine forms of biopower where nonhumans, particularly some animals, are objectified as data producers [28].

Digital ecologies framework informs design on directions and perspectives for ethical use and effective, positive consequences of technologies for ecological purposes. It influences and enlightens our research scope and epistemology: We aim at conviviality and reciprocity rather than unilateral user experience and promote materialities as the manifestation of invisible dynamics and biopolitics.

3.2.4 Ecological Design

What is the shape of natural ecosystems? How can we reveal the function of ecosystems from their shape? Can we infer and use ecological networks to integrate our human-made design to maximize ecosystems' function? These questions are profoundly relevant, the more climate extremes affect us yet force us to redefine design in a consequential way, to control climate from the indoor to the outdoor, which has positive cascading implications for all species in ecosystems. Then how can we combine beauty (aesthetic qualities driven by human intentionality that is the leitmotiv of design) with function (so dear to engineers regardless of the aesthetics)? The answer is in nature.

Ecology, defined as the organization of ecosystems' components (species and "abiotic" elements), constitutes the "house" (natural and artificial) of ecosystems: a definition deeply rooted in ecology's etymology. The design of nature (ecology as interactions) is driven by unconscious evolutionary mechanisms that optimize function (or equivalently minimize energy consumption) through the rewiring of networks under environmental pressure [29, 30]. This optimality is centered on integration, connectivity, and adaptation, which should be crucial for conscious human-made design and, more importantly, in integrated ecosystem development, where conscious design is the core of the connection between the natural and built environment. In this type of novel design, three elements are sought to be crucial for ecosystems:

(i) ***Eco-inspired***, as topologically similar to nature's interaction networks. Ecosystems organization has a variety of degrees of order that can be quantitively assessed through the assessment of interactions' topology, resulting in emergent function, where for the latter, the most important is hydroclimate regulation (water cycle and temperature). The optimal topology is the Pareto scale-free network configuration [31], where a tree-like structure is upscaling all local functions to the system level (Fig. 3.1). In this sense, an eco-inspired design must mimic nature's scale-free organization for optimal function at the community [32, 33], building, and any other scale the design is reaching [30, 34].

(ii) ***Eco-integrated***, as coalescent structures where the type of ecological components reflect the neighboring structures or environmental fields. In a forest, species integrate each other (mirroring or strikingly diverging from others) so that their nexus leads to self-regulating ecological flows, such as invisible food-web regulations strongly linked to the ecological structural networks. In this regard, design can and should incorporate species that mirror, complement, and attract species found in the natural environment.

(iii) ***Eco-reactive***, as structures where natural eco-inspired and eco-integrated design is an adaptive and enhancing function. This eco-reactivity can be real-

Fig. 3.1 Multiscale ecosystem structure. From left to right, habitat, species, and subspecies networks reveal nature's hidden and visible self-organization dictated by interacting components that shape habitats. A systemic scale-free network, that is, the hydrologic network, is responsible for the water provision and species dispersal in forests, yet for the organization of trees (left plot); the scale-free vascular network for the supply of nutrients to capillaries (middle); and the looping regular network of capillaries in all leaves for maximizing their photosynthetic capacity (right plot). Scale: 500–0.1—0.05 meters (Image produced by the author M. Convertino with original pictures taken in Dapeng, Shenzhen)

> ized by combining different network structures at different scales (in analogy to the leaf as in Fig. 3.1) or by strengthening bio-terraforming species function (e.g., by enhancing their spatial spreading and form). Bioterraforming species are critical because they form habitats and are incredibly eco-functional, such as local keystone plants (mosses and ferns) along gradients of structures' variability. In this purview, design can embrace the difficult but charming intentionality that is similar to the creation of novel material with superior performance: an exercise that is likely played out in nature with the introduction of novel species or the extensive adaptation of existing ones under strong environmental pressure such as heatwaves and blooms.

Nature is eco-inspired, eco-integrated, and eco-reactive. By design, it is possible, and we argue necessary, to be eco-inspired, eco-integrated, and eco-reactive (in general, *eco-based*), particularly considering the challenges dictated by climate change, sustainability of natural resources, and the overall survivability of the planet. The patterns we observe in nature are symptoms of function, anomalies, and opportunities to act on underpinning interactions defining patterns. Design can shape the intended structures to be eco-functional toward the priority goals we set. These design principles are valid across scales, from eco-inspired or eco-based technology [35], garments, buildings, and communities to communication—as information that is routed correctly with ecological meaning [36]—and innovative artifacts, where beauty is a sign of optimal ecological organization and function. Ecophilia should not be just the human instinct to be attracted to secluded nature but the intention to live with nature.

3.3 Cases and Discussion: Ecological Materialities

This paragraph uses our research projects to examine the theoretical framework outlined above. The cases are ongoing explorations of materialities presented through their conceptual, applicative, and semantic value in ecological design.

We discuss (1) the value of ecological information and functions for materials enhancement; (2) the use of digital technologies and data for encounters other than humans and their comprehension; and (3) the adoption of cross-disciplinary, "contaminated and undisciplined" methods.

3.3.1 Ecology-Informed Materials

How does ecological design inform material properties, functions, and identities of future habitats?

The authors' collaboration starts from a situated setting, the high humid conditions of the South of China, where they live and work.

Walls are a crucial built habitat component, serving as structure and partition while protecting from external conditions such as unfavorable and hostile natural environments. However, the traditional industrial materials used for constructing and covering walls adopt the same property in any geographical area, climate, type of indoor environment, and human beings: they are impervious. This means that the main property transferred to the interfaces between us, our living space, and the environment is to be waterproof and anti-microbial.

We propose a new identity for our indoor living surface: porous breathing skin that can self-regulate the humidity condition, temperature, and microbiome.

Breath aims to optimize indoor ceramics based on ecological function, information, and simulation [37]. It is a breathing material system for enhancing the quality of the indoor environment through the design of material devices that control humidity, temperature, and microbial conditions with no or limited energy use. The strategy is to design tailored porosity across material scales, from the nano porosity of the ceramic ingredients to the mesoscale of tiles' geometries and the macro, ecological scale of material infrastructures, which guide the moisture around the space based on physical principles. The natural organization and texture distribution inspire the structural porosity obtained by computation design (Fig. 3.2) to improve biomimetic materials and design principles. A deeper understanding of the effects of ecological networks' autocorrelation structure and texture distribution can help achieve more precise biomimicry through strategies of heterogeneous systems.

Responsive surfaces can be optimized using porous, networked structures that auto-regulate based on contextual conditions. The study has generated several porosity-networked structures-texture combinations that are being 3D printed and their properties tested in controlled and actual environments.

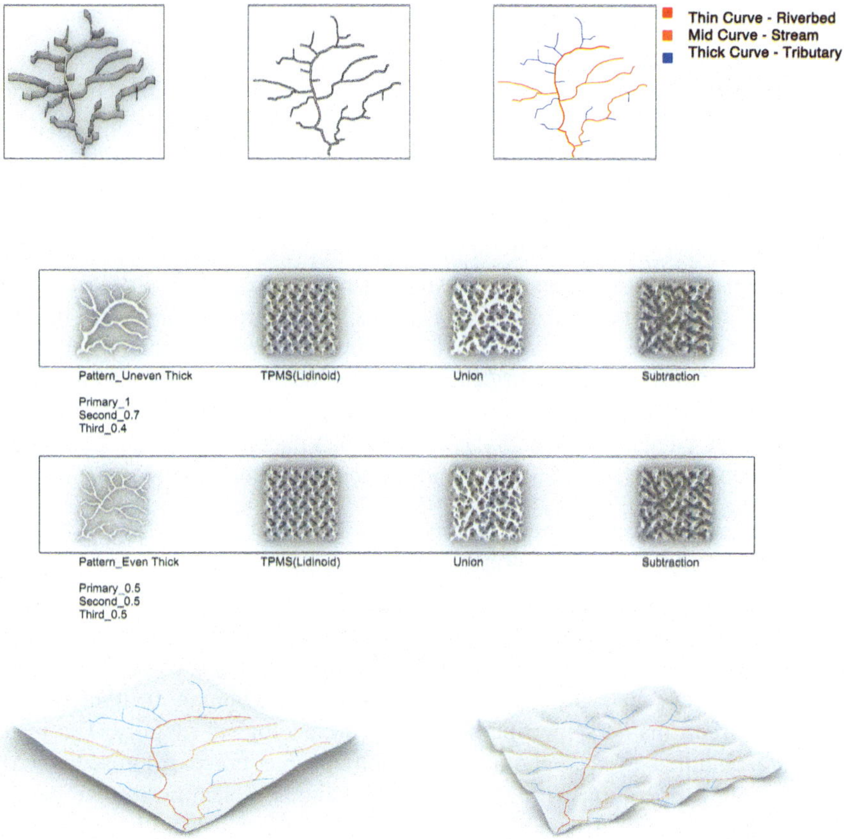

Fig. 3.2 Methodological scheme for synergizing porous structures obtained by computational design with networked geometries for innovative ecological ceramic systems with enhanced hygroscopic properties. The original image was produced by students YAO Zhennan and CHEN Yinglei of the Materialities Research Group, SUSTech School of Design, under the guidance of the author E. Migliore

To expand our knowledge of bio-digital systems' suitability for indoor environments, we will regularly analyze the microbiome developing on ceramics and discuss models of positive, caring, and advantageous exchange and coexistence with entangled materialities.

Innovating indoor materials involves introducing the concept of porous, ecological networks into the design. Traditional material design often overlooks the influence of spatial dimensions on the material's overall function by focusing on a single or evenly distributed attribute. However, natural ecosystems exhibit a rich functional network structure, which can be used to create well-designed and adaptable materials. By simulating the texture distribution of ecological networks and colonizing them with living organisms, we can optimize the function of materials while promoting and appreciating the value of human-nonhuman conviviality.

3.3.2 Technology-Enabled Material Ecologies

How do digital technologies enable understanding and simulation of ecological functions and identities? Which are these technologies in research and practice?

Technologies for ecology are a research interest of the Materialities Research Group at the SUSTech School of Design in Shenzhen. Our research adopts several types of techs in applied and speculative projects. Advanced microscopical analysis allows the observation and study of previously impossible materials, including unicellular organisms and nanometric particles. Sensors enable real-time, accurate data collection in space and time, providing crucial information on materials, behaviors, changes, and responses to conditions. 3D scanning technologies empower the faithful digitization of complex, organic forms in the virtual space. Computational Design helps create digital models of sophisticated geometries and uses algorithms to manipulate them parametrically. Advanced manufacturing technologies allow the three-dimensional rematerialization and the scalability of simulated natural geometries in the real world.

The project *A Grain Of* [38] (Figs. 3.3 and 3.4) explores urban ecology through sensing, observation, and sound recording technologies. It adopts 3D printing manufacturing to translate the data into tangible, material interfaces for beyond-the-screen encounters. Migliore et al. [38] collected and recorded the situated sound of soil and water samples from Shenzhen, where they live and work. The goal is to compile data that give access to unsettled other-than-human encounters to help draw the urban matter's invisible and inaudible ecological identity. The material data are obtained in the laboratory through Atomic Force Microscopy (AFM) (Fig. 3.3) in cooperation with Massimiliano Galluzzi, Scientific Researcher at the Shenzhen Institute of Advanced Technology. Data are translated into 3D morphological maps and printed to become surfaces for human interface. The authors

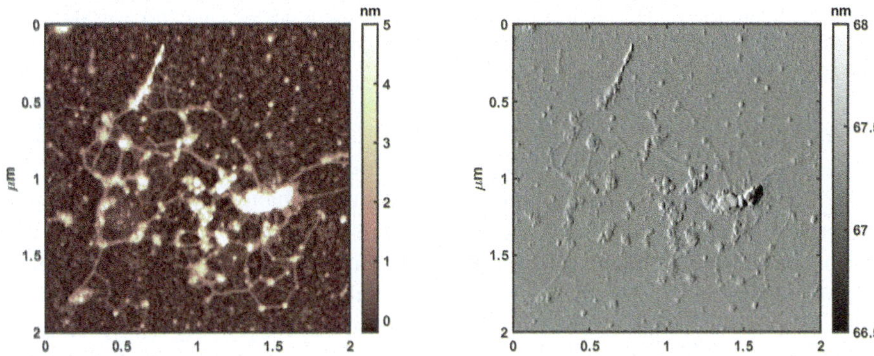

Fig. 3.3 Analysis of urban soil and water through Atomic Force Microscopy (AFM). (Original images produced by Massimiliano Galluzzi, SIAT, for the project Migliore E, Sagesser MZ, Qu H (2023) A Grain of: A Multimodal Representation of the City's Ecology Through Material and Sound. In: ACM International Conference Proceeding Series)

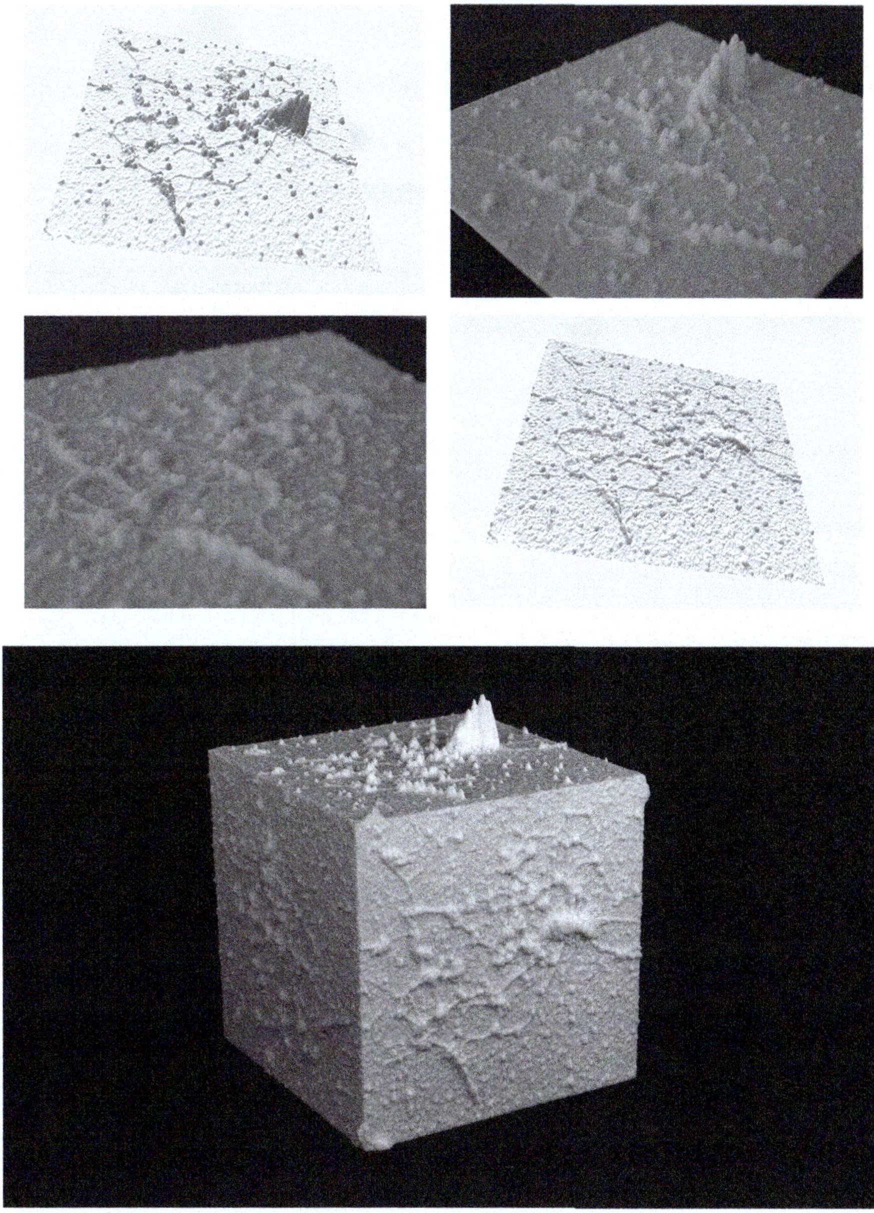

Fig. 3.4 Materialization of microscopic data of soil and water samples from Shenzhen City. (The original pictures show the project Migliore E, Sagesser MZ, Qu H (2023) A Grain of: A Multimodal Representation of the City's Ecology Through Material and Sound. In: ACM International Conference Proceeding Series)

Fig. 3.5 3D scanned models of materials collected on the ground of the historical Hakka village in Pingshan, Shenzhen, Guangdong province in China. On their surface, moss, organisms, and soil become an inextricable part of their virtual materiality (Original image produced by the author E. Migliore from digital models obtained by students of Materialities Research Group)

presented the first step of this project at the Seventeenth International Conference on Tangible, Embedded, and Embodied Interaction [38] in the form of a cube 25×25×25 cm emitting sound from within: It is supposed to represent a grain of the city and become a scalable and replicable model to be filled with contents and data from different places and different cities.

The project *Future Archaeologies* is speculative research about the future of our material remains and the adoption of found materials as interfaces for historical, cultural, and natural encounters. The researchers 3D scanned, through Shining 3D Einscan SP, 21 materials located on the ground of the historical site of Hakka village (Fig. 3.5) in Pingshan, Shenzhen, Guangdong province in China. The digital archive of these objects freezes a set of taxonomic data in time and shapes. It allows the digitization of nature thanks to the high resolution of different granularity. Most of the samples show an entanglement of inert and organic materials. Once frozen in the digital space, moss, soil, and organisms become part of the identity of that materiality, its texture, and its functional and aesthetic power.

3.3.3 Cross-Disciplinary Research and Undisciplined Methods

Our research involves collaboration among various disciplines, including product design, environmental engineering, structural engineering, sound design, media art, materials science, archeology, and architecture. Tony Dunne, Professor of Interaction Design at the Royal College of Art, London, states: "New hybrids of design are emerging. People do not fit in neat categories; they are a mixture of artists, engineers, designers, thinkers" [39].

By working together across different fields, we aim to facilitate the exchange of knowledge and technological innovations and provide comprehensive opportunities and perspectives to address ecological and environmental issues without disconnecting them from culture and society. The complexity and entanglement of nature and design should be reflected in a likewise heterogeneous disciplinary team. The shift from an anthropocentric to a nonhuman inclusive perspective cannot be sustained and advanced by existing disciplinary separations.

Deleuze and Guattari [40] advocate for a new kind of science that prioritizes ethical transformation and is not constrained by the economic demands of advanced capitalism. This new science is called "minor science" and is practically oriented. The focus is on providing local knowledge to achieve specific goals while acknowledging a dynamic and heterogeneous world that could be more stable and consistent. This approach incorporates experiential and corporeal sensing and values affective processes before consciousness and reflection. This minor science aims not to represent the world but to generate difference, divergence, and creation. In this framework, it is evident that creative disciplines like design need a substantial shift toward a multi-, cross-, inter-, and even "alter"-disciplinary approach [41]. Rodgers et al. [41] provide a series of disciplinary—or, using their word, 'alterplinary'—attributes and agencies of the future design methodology. Notably, our research resonates with their idea of tackling the unknown as the motivation for radical innovation, the encouragement to take design seriously, being a strategic and competent actor in complex scenarios, rejecting the idea of sustainability without responsibility, inviting design to become 'undisciplined,' which means "[...] to shift methods from being 'discipline-based' to 'issue or project-based.' These designers will be most effectively placed to make connections that generate new methods and to identify 'other' dimensions of design research, activity, and thought needed for the complex, interdependent societal issues we now face" [41].

This essay is itself a demonstration of the authors' cross-disciplinary and situated methods. They are operatively collaborating on the project *Breath*, described in Sect. 3.6.1, and intellectually, ethically, and in an "undisciplined way," hybridizing their disciplinary perspectives and scopes.

All the projects described are situated and carried on with an assemblage of methods to avoid aggregate data to produce uniformity and underplay real-world changes [4]. What is clear amidst this complexity is that these are "compromised times" where notions of "purity" must be abandoned [42].

3.4 Conclusions

Designing today for the future is an exciting and complex challenge requiring a new setting of lenses, approaches, and methods. The planetarian environmental crisis dismantled our faith in human progress and certainties. As creative actors, what do we need to know and do, and how to achieve it to contribute to a favorable future?

In this chapter, we corroborate theoretical and operative frameworks guiding the design community in facing the challenge of complexity. We also bring our activities as cases for discussing possibilities, alternatives, and applications.

We embrace scholarly post-anthropocentric theories and approaches aimed at emancipating the affective capacities of the nonhuman, while championing their functional role in an ecological design.

Biodesign is proposed as the cross-disciplinary framework of emerging practices of codesigning with nature, whose principal value is to favor a narrative of coexistence, respect, and gratefulness. New materialism is the philosophical perspective on human-nonhuman entanglement that offers the theoretical background to subvert disciplinary and intellectual attitudes. It draws on emerging physical theories of the matter and defines materialities as plural, open, complex, uneven, and contingent. It also helps conciliate social and natural sciences, thus facilitating cross-disciplinary dialogues. Digital ecologies is the analytical framework for exploring the digital entanglement in nature ontologically and ethically. It offers a lucid analysis of a very knotted phenomenon to guide design decisions and directions ethically and critically. Our research benefits from its material perspectives on digital ecologies and derives a clear purpose of conviviality and reciprocity, taking distance from design for spectacular, generalized, and unilateral user experience. Ecological design is the proposed operative framework for ecology-informed design. We claim that to maximize ecosystem function, the design outputs should be *Eco-inspired*, as topologically like nature's interaction networks; *Eco-integrated*, as coalescent structures where the type of ecological components reflect the neighboring structures or environmental fields; and *Eco-reactive*, as structures where natural eco-inspired and integrated design is adaptive and enhancing function. Finally, we propose ecological materialities as future materials design's ontological and methodological identity. It is (1) ecology-informed as it will imitate properties and behaviors of ecological reciprocity instead of being rigid, homogenous, passive entities; (2) technology-enabled, since digital tools and data allow unprecedented encounters and intimacy with nonhumans and empower their role in design practices, and (3) undisciplined, shifting from being "discipline-based" to "issue or project-based." Cross-disciplinary and hybrid epistemology enables the design to comprehend and respond to real-world changes, where disciplines need to renounce the idea of purity and uniformity to navigate the complexity.

Acknowledgments We thank the students CHE Ran, YAO Zhennan, and CHEN Yinglei for their materials and computational design research. A special thanks to Massimiliano Galluzzi, Scientific Researcher at the Shenzhen Institute of Advanced Technology, Chinese Academy of Science, for his support and ideas when creating microscopic analyses of our material samples in the lab.

References

1. Capra F (2007) Sustainable living, ecological literacy, and the breath of life. Can J Envir Educ 12(1):9–18
2. Orr DW (2002) The nature of design. Ecology, culture, and human intention. Oxford University Press, New York
3. Bateson G (1972) Steps to an ecology of mind. Ballantine Books, New York
4. Fox NJ, Alldred P (2019) New materialism. In: Atkinson PA, Delamont S, Cernat A et al (eds) SAGE Research Methods Foundations. SAGE, London
5. Braidotti R (2006) Posthuman, all too human: towards a new process ontology. Theory Cult Soc 23:197–208. https://doi.org/10.1177/0263276406069232
6. Myers W (2018) Bio design. nature+science+creativity, July 24, 2018. The Museum of Modern Art, New York
7. Collet C (2013) Alive: new design frontiers. Exhibition curated by Carole Collet April–September 2013, Espace Fondation EDF, Paris. www.thisisalive.com
8. Grushkin D, Vijayakumar V, Eastman C (2022) Grow the future: visions of biodesign. Biodesign Challenge, New York
9. Mundt K (2020) Decoloniality as praxis. Cult Stud 34:489–492. https://doi.org/10.1080/09502386.2019.1597140
10. Deleuze G, Guattari F (1984) Anti-Oedipus. Capitalism and schizophrenia. Athlone, London
11. Barad K (1996) Meeting the universe halfway: realism and social constructivism without contradiction. In: Feminism, science and the philosophy of science. Kluwer, Dordrecht, pp 161–194
12. Braidotti R (2006) All too human: towards a new process ontology. Theory Cult Soc 23:197–208
13. Barad K (2007) Meeting the universe halfway. Quantum physics and the entanglement of matter and meaning. Duke University Press, Durham
14. Coole DH, Frost S (2010) New materialisms. Ontology, agency, and politics. Duke University Press, London
15. Elvin K, Bahareh B, Elisa G (2020) Living artefacts: conceptualizing livingness as a material quality in everyday artefacts. Int J Des 14:37–53
16. Braidotti R (2013) The posthuman. Polity, Cambridge
17. Turnbull J, Searle A, Hartman Davies O et al (2023) Digital ecologies: materialities, encounters, governance. ProgEnviron Geogr 2:3–32. https://doi.org/10.1177/27539687221145698
18. Searle A, Turnbull J, Hartman Davies O et al (2023) Glitches in the technocultural present. Dialogues Hum Geogr
19. Leszczynski A (2015) Spatial media/tion. Prog Human Geogr 39:729–751
20. Balmford A, Cowling R (2006) Fusion or failure? The future of conservation biology. Conserv Biol 20:692–695
21. Louv R (2005) Last child in the woods. Algonquin Books, Chapel Hill
22. Kareiva P (2008) Ominous trends in nature recreation. Proc Natl Acad Sci USA 105:2757–2758
23. Pyle RM (1993) The thunder tree: lessons from an urban wildland. Houghton Mifflin, Boston
24. Kuntsman A, Rattle I (2019) Towards a paradigmatic shift in sustainability studies: a systematic review of peer-reviewed literature and future agenda setting to consider environmental (un)sustainability of digital communication. Environ Commun 13:567–581
25. Sy T (2016) Technofossils of the anthropocene: media, geology, and plastics. Cult Politics 12:355–375
26. Taffel S (2021) Data and oil: metaphor, materiality, and metabolic rifts. New Media Soc 25(5):980–998
27. Ash J (2009) Emerging spatialities of the screen: video games and the reconfiguration of spatial awareness. Environ Plan 41:2105–2124
28. Buscher B (2016) Nature 2.0: exploring and theorizing the links between new media and nature conservation. New Media Soc 18:726–743
29. Bak P (1996) How Nature Works: The Science of Self-Organized Criticality. Springer, New York. https://doi.org/10.1007/978-1-4757-5426-1

30. Banavar JR et al (1999) Size and form in efficient transportation networks. Nature 399(6732):130–132
31. Mandelbrot B (1982) The fractal geometry of nature. W. H. Freeman and Co., San Francisco
32. Galbraith E, Convertino M (2021) The eco-evo mandala: simplifying bacterioplankton complexity into ecohealth signatures. Entropy 23:1471
33. Galbraith E, Frade P, Convertino M (2022) Metabolic shifts of oceans: summoning bacterial interactions. Ecol Indic 138:108871
34. Convertino M, Valverde JL (2019) Toward a pluralistic conception of resilience. Ecol Indic 107:105510
35. McLamore ES, Convertino M, Ocsoy I et al (2017) Biomimetic fractal nanometals as a transducer layer in electrochemical biosensing. In: Semiconductor-based sensors. World Scientific, pp 35–67
36. Servadio J, Convertino M (2018) Optimal information networks: Application for data-driven integrated health in populations, vol 4. Sci Adv, p e1701088
37. Migliore E, Yao Z, Deng X (2023) Porous ceramics for the design of domestic ecologies. In: Marcus A, Rosenzweig E, Soares MM (eds) Design, user experience, and usability, HCII 2023. Lecture notes in computer science, pp 436–453
38. Migliore E, Sagesser MZ, Qu H (2023) A Grain of. Proceedings of the seventeenth international conference on tangible, embedded, and embodied interaction. ACM, New York, pp 1–4
39. West D (2007) A new generation. ICON 43:56–64
40. Deleuze G, Guattari F, A thousand plateaus
41. Rodgers PA, Bremner C (2017) The concept of the design discipline. Dialectic 1:19. https://doi.org/10.3998/dialectic.14932326.0001.104
42. Shotwell A (2016) Against purity: living ethically in compromised times. University of Minnesota Press, Minneapolis

Enza Migliore is an assistant professor at SUSTech School of Design as the PI of the Materialities Research Group. She is a design researcher, practitioner, and educator dedicated to materials exploration through interdisciplinary collaboration and experimental practices. Through her design research, she informs scientific laboratories with systemic knowledge, creative practices, and opportunities.

Matteo Convertino is an associate professor at Tsinghua SIGS as the PI of the TREES lab. His expertise is in ecological intelligence, biocomplexity (patterns, networks and ecostress), ecohydrological risk synthesis and forecasting, ecosystem decision making, eco-design and engineering. He has published over 120 publications, including 86 in Q1 journals in Ecology, Environmental Sciences, Risk and Engineering. He was among the founders of the EcoHealth Division at the University of Minnesota Twin Cities (USA) and the Ecological Big-Data group at Hokkaido University (JP).

Chapter 4
Nature Sparkles by Design: Natural Molecules for Polymer Functionalization for Sustainable Design Applications

Francesca Nicolais, Angela Longo, Mariano Palomba, Loredana Schiavo, and Gianfranco Carotenuto

Abstract In nature, there is a lot of light that we are unable to see that enables and reveals the communication among invertebrates, fish and reptiles, amphibians, and birds. This is the invisible bio-fluorescence phenomenon. Recently, rosin, a yellow, solid, transparent plant resin is being investigated because of its unique luminescence. A strong physical interaction between polymer and rosin is possible and such property causes the formation of a solid solution, which is optically transparent and perfectly suitable for a large number of sparkling design applications that are more sustainable and in harmony with a new way of thinking and designing that respects man and the environment. Great interest is in the design of artifacts aimed at coworking spaces, with a biophilic design approach and inspired by biomimicry, in which surfaces and plants are illuminated even at night, providing acoustic and visual protection while generating atmospheres of well-being.

Keywords Rosin · Natural functional molecules · Luminescence · Solid-solution · Design · Biomimicry

4.1 Introduction

Materials science has moved on from a phase of materials with controlled complexity in which one had to produce perfect materials, to one in which anisotropies and additive doping are specially designed to achieve specific performance that makes

F. Nicolais (✉)
Università degli Studi Suor Orsola Benincasa, Naples, Italy
e-mail: francesca.nicolais@unisob.na.it

A. Longo · M. Palomba · L. Schiavo · G. Carotenuto
Institute for Polymers, Composites, and Biomaterials, National Research Council, (IPCB-CNR) UOS Napoli/Portici, Portici, NA, Italy
e-mail: angela.longo@cnr.it

A. J. V. de Arruda, F. L. Palombini (eds.), *Biology, Biomimetics and Natural Design*, Environmental Footprints and Eco-design of Products and Processes,
https://doi.org/10.1007/978-3-031-65877-8_4

advanced materials mimic natural organic high-functionality materials, able to solve highly engineered problems by producing complex structures that often mimic the micro-morphologies of natural structures.

Today, the properties of a material can be custom-designed by manipulating its structure at different size scales, from the macroscopic to the atomic. Matter, as if drawing inspiration directly from nature, is becoming more and more like a real living organism capable of feeling and reacting to stimuli and developing systems based on the properties (structure, function processes, organization, and relationships) of biological systems [1]. It thus gives rise to artifacts that replicate the structural and formal functional characteristics of natural systems by studying biological and biomechanical processes as a source of inspiration [2] or that are built upon the logics and principles that underpin the organization and evolution of biological species by translating them into the form of codes and algorithms. One such example is "parametric biomimetics" [3], which fosters the achievement of morphologies that reproduce "those principles of structural optimization of hierarchisation, energy-saving, resilience, synergetic collaboration, adaptability, and flexibility that in nature underlie the production of such morphologies and can prove valuable in design" [4].

Biomimetic imitation of natural models, or "learning by nature" [5], can be adopted in surface or structural design and has great application potential for materials science and in particular composite materials. The latter have been widely used in multiple engineering applications due to the unique combinations of form and mechanical and physical properties that only they can achieve [6].

Nature-inspired materials are developed in order to exploit a certain type of property, which allows us to obtain a particular type of benefit that may be, by way of example, but not of limitation, (i) electrical, (ii) biological, (iii) chemical, (iv) mechanical, (v) sustainable, (vi) or be related to a combination of such benefits [7] that can sometimes even exceed nature.

The materials designer can draw inspiration and support from the biological sciences by developing concepts and solutions, which, going beyond formal and aesthetic characteristics, can express the innovative potential that natural models can offer [8]. In this scenario, relations between the logics, principles, structures, and strategies observed in nature become creative triggers for the production of innovative-engineered materials that mimic nature.

One of the best-known examples of bioinspired imitation of natural models for the creation of new materials with peculiar performances is Velcro, a fastening device inspired by burdock, a plant organism. Created in 1948 by the Swiss engineer Georges de Mestral, it imitates Burdock's strategy for attachment to surfaces via its peculiar hook-like structure. Other examples are nonslip shoes with grooves similar to the pads of dogs' paws; swimwear inspired by shark skin [9]; water-repellent, self-cleaning fabrics or building façades that make use of the lotus effect (lotus leaf-inspired self-cleaning materials) [10], whereby water droplets collect dirt as they run downwards, keeping the leaf surface clean. Lotus leaves in fact have special micro-rough surfaces, which greatly reduces the contact surface area. So, when a drop of water falls on the leaf it remains a drop and retains its spherical shape. This way it easily rolls down with the help of gravity or the wind, collecting dirt as it goes. In some cases, nature itself offers useful and readily usable materials for the development of new technological applications. Zeolites, for example, are a class of very smart natural materials having a nanostructured inner texture. They are microporous three-dimensional crystalline solids of aluminum silicate characterized by the presence of regular microporosity and a characteristic crystalline structure, with a fibrous, lamellar, or globular texture [11, 12]. Among them, the natural clinoptilolite, which is one of the most common natural zeolites, is characterized by extremely compact stacks of single lamellar crystals with a thickness of 40 nm and the other two dimensions of several tens of microns [13]. Such a special type of lamellar texture, clearly visible in Fig. 4.1-up, has allowed to produce a nanopowder made of 40 nm-thick nanoplatelets simply by exfoliation of the geomorphic mineral [14]. The result of the exfoliation treatment is shown in Fig. 4.1.

Such nanopowder can have different technological applications, for example, molecular traps for VOCs [15, 16], very active drying agents [17], gas-separation process [18], ion exchange materials [19], etc.

The replication of natural strategies and the transformation of organic materials into ultramaterials is both an aspiration and a challenge for materials designers; think, for example, of how living matter can self-structure down to atomic levels; or how nanotechnological research seeks to reproduce photosynthesis by imagining an artificial photosynthesis, or nature-inspired photosynthesis [20], with the goal of unlimited energy resources. Other possibilities are materials with color variations inspired by birds' wings or peacock feathers; the bio-fluorescence strategies found in a range of nocturnal-crepuscular organisms including invertebrates, birds, and mammals [20]; the skin of cephalopods as a model for creating a flexible surface capable of producing different patterns of fluorescent light [21]; the hypothetical possibility of lighting trees by experimenting with injections of gold nanoparticles into the leaves of the aquatic plant Bacopa caroliniana to stimulate the chlorophyll in the leaves to emit red light [22].

In some cases, living matter in fact presents limited possibilities, and nanotechnology tries to push and extend natural solutions, making it possible to create extreme artificial conditions in terms of vacuum, purity, and cold. By thus revealing the astonishing properties of matter at nanoscopic level, it becomes possible to make choices at a molecular level by performing increasingly targeted and microscopic actions on the matter, leading to the discovery of vistas of materiality.

A time in which "nature enters the design project not only procedurally as inspiration or reference but is hosted and extended by the project as code and organic material for its life. Nature becomes a mechanism and is modified by man; it seems to become the programmed host of bodies synthesized by its own creations" [23]. Nanotechnology—that is, technologies that allow matter to be manipulated and controlled at atomic and molecular levels—is one of the most important developments in materials science today and is revolutionizing product design by making it possible to build materials, devices, and systems by controlling matter at billionths of a meter scale and to endow materials with structural and functional properties that have never been imagined previously. These materials with new performances

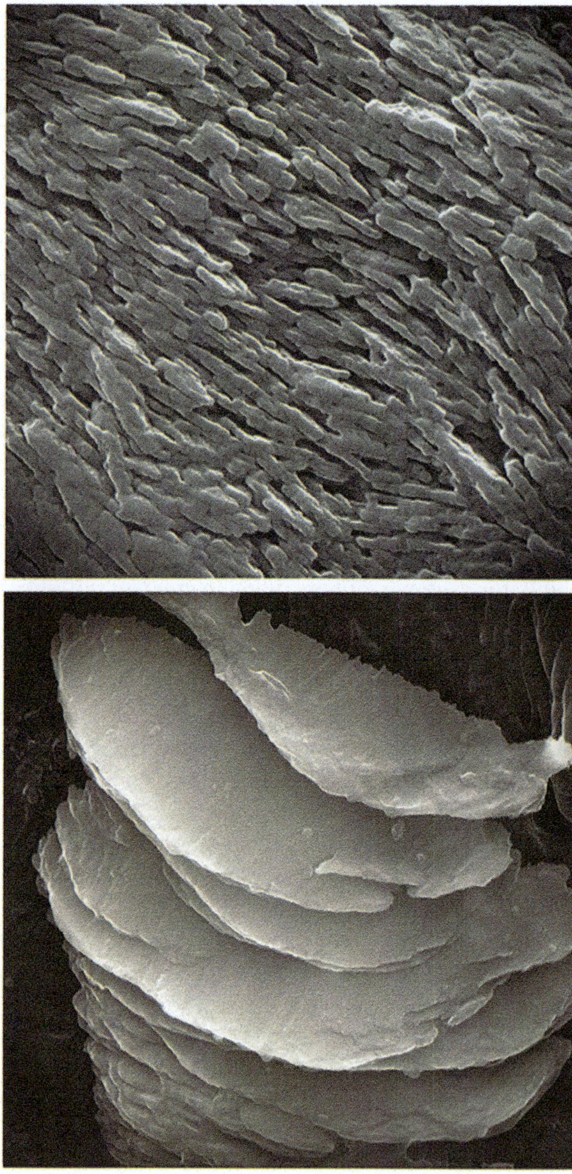

Fig. 4.1 SEM-micrographs of natural clinoptilolite lamellas before (up) and after (down) the exfoliation treatment. ** *The SEM-micrographs are part of the photographic exhibition "Progettare l'invisibile: natura e tecnologie per il design delle nanostrutture". Curated by Loredana Schiavo, Angela Longo, Mariano Palomba, Francesca Nicolais, Maria Cristina Del Barone e Carotenuto Gianfranco. In collaboration with Lorena Affatato and the Riot studio. The Photographic exhibition is among the events of the centennial celebrations of the National Research Council CNR (1923–2023)*

increasingly resemble "augmented" biological organisms that allow a physical and emotional relationship to develop with the user, thereby enhancing the senses and generating new interactive modes.

An excellent example of virtuous imitation and implementation of nature are "smart materials" [24], which respond to the application of an external force field by modifying one or more of their properties (mechanical, optical, electrical, magnetic, chemical, or thermal) and adapting to changes in the environment. This way they feel, act, and learn from the environment. They are complex performance systems composed of different materials. Smart materials are made up mainly of actuators, which can alter their characteristics (shape, thickness, stiffness, position, etc.) as environmental conditions change, or sensors, which provide the actuators with information generated by structural and environmental changes. They are intelligent materials with new functionalities and the ability to change their shape and to self-heal.

An effective approach for the development of smart materials involves nanocomposites based on the functionalization of plastics by means of nanostructures [25–28].

Nanocomposites are artificial polyphasic solids, meaning that they are obtained by generating different solid phases through synthesis [29, 30]. The dispersed solid phase is called the filler, while the dispersing solid phase is known as the matrix. The distinction with ordinary composites lies in the size of the filler, which in nanocomposites must be nanometric (a nanometer is a billionth of a meter, i.e., 10–9 m). Certain properties are found in solids only when their size is reduced to a few nanometers or even a few Angstroms (an Angstrom is a tenth of a nanometer, i.e., 10–10 m), yet certain characteristics emerge in matter even when it is several tens of nanometers in size. These solids, which are neither massive nor made up of individual atoms or molecules, have properties that are said to be mesoscopic and are absolutely peculiar to these systems. These new properties originate mainly from surface and electronic overrun effects and are often very useful in various areas of technology. The filler in the matrix (nanocomposite topology) may have a regular or irregular spatial distribution and the individual particles can have a very varied morphology (spherical, needle-shaped, cubic, etc.). In these systems, control of morphology is a decisive aspect, as the mesoscopic properties are closely dependent on it. Since the filler dispersed in the matrix is almost at an atomic level, nanocomposites should be rightly considered homogeneous systems (unlike the corresponding macro- and micro-composites) and therefore made up of true alloys or solid solutions of extremely incompatible phases. For example, metal-polymer nanocomposites are actually solutions of metal in polymer or metal-polymer alloys. They, therefore, constitute a completely new, broad class of materials with absolutely unique properties.

Polymer matrix nanocomposites have numerous possible technological applications. For example, by dispersing magnetic nanostructures (e.g., iron oxide nanoparticles) it is possible to obtain magnetic plastics that are still light and transparent, making them suitable for use in the microwave sector, for example [31]. Conductive plastics can be engineered by dispersing conductive nanostructures such as graphene or reduced graphene oxide nanoplatelets. These nanostructures can conduct electric current because of the presence of delocalized molecular orbitals above and below the molecular plane. A polymer matrix such as silicone can therefore be made conductive by applying a small amount of these nanostructures These particular systems are seen to possess an electrical resistivity that varies with the application of pressure or deformation, so they lend themselves to the production of various functional electrical devices (e.g., switches, transducers, etc.) [works involving graphene].

In addition, the nanometric particles of some metals such as gold, silver, and copper are characterized by strong absorption in the visible region—a phenomenon known as plasma resonance. These particles have long been used for staining church windows: Their small size renders the glass medium completely transparent to sunlight while also being intensely colored. The metal nanoparticles have bright, intense coloration. In particular, silver nanoparticles have a straw-yellow color; gold nanoparticles are ruby-red, and copper nanoparticles are aqua-green. A fundamental characteristic of these pigments is their absolute stability in sunlight. And since this coloration is an intrinsic property of the metallic crystal, and remains completely unaltered even after prolonged exposure to intense sunlight, the original coloration remains unchanged even after decades. Polymers can also be colored via the incorporation of gold, silver, and copper metallic nanoparticles. The colors of these materials are so intense and brilliant that they can be used to make ultra-thin optical filters. In addition, any metal nanoparticle aggregates in the polymer enable the color to vary with temperature, a phenomenon called thermochromism. This is due to the sensitive thermal expansion of polymers at temperatures above the glass transition temperature (Tg). In fact, at above-Tg temperatures, the matrix encapsulating the nanoparticles will expand significantly, causing the particles to recede significantly and thus allowing the transition from coloring by interacting nanoparticles to that more characteristic of isolated nanoparticles. Polymers filled with aggregated metal nanoparticles therefore lend themselves to the manufacture of thermochromic plastics. Such materials will be capable of changing from brown to yellow when based on silver nanoparticles, and from violet to red when made up of gold nanoparticles. The temperature at which this color change takes place depends on the glass temperature of the polymer [32].

In addition to possessing the property of plasmonic resonance, nanostructures of precious metals such as gold and silver, when further reduced by a few nanometers through the adoption of semiconductive behavior, can acquire luminescence. These nanostructures become functional for the engineering of transparent and luminescent optical plastics. In fact, it is possible to avoid opacification of the polymer by introducing a filler of an average size that is sufficiently smaller than the wavelength of light radiation [33, 34].

An innovative technology in the field of anti-counterfeiting [35] uses this new class of nanostructured fluorescent additives for thermoplastic polymers. It makes them highly luminescent on exposure to ultraviolet light and tunable by adjustment of the chromatic hue of the colored light exposed to ultraviolet radiation of an appropriate frequency by controlling the nature of these additives and their characteristic size. Moreover, these nanocomposites, leaving the properties of the polymer unaltered, represent an extremely versatile solution due to their processability and offer great flexibility in numerous fields of application.

Recently, also in the area of functionalizing plastics, there is an emerging scenario geared towards nature, in terms of sustainability—linked to the life cycle of materials—and in terms of nature as a model in accordance with a biomimetic approach to innovation. It would appear then that the study and use of natural organic molecules characterized by peculiar optical properties such as luminescence, etc., is becoming increasingly interesting and beneficial. The results of this approach, which is based on the preparation of solid solutions, will be discussed and illustrated with a case study in the following paragraphs.

4.2 Materials Based on Solid Solutions of Functional Molecules

A variety of new industrially useful materials has been developed by combining (Latins would say: 'cum grano salis') some of the already existing substances like, for example, polymers, resins, lapideous materials, organic molecules, coordination complexes, organometallics, inorganic/organic salts, biopolymers, enzymes, vitamins, natural substances, and many other types of high and low molecular weight chemical compounds. This "trial-and-error" approach has been widely exploited in materials science and composites could be considered the most representative example, since they are multiphasic solid materials resulting from the combination of at least two different solid phases: (i) a continuous solid matrix and (ii) a dispersed solid phase, known as filler or reinforcement. In particular, common composites have been obtained by combining polymeric matrices (both thermoplastics and thermosetting resins) with ceramics (carbon, glass, etc.) or metals, shaped in the form of particulates, short/long fibers, platelets, etc. The most technologically used combinations are: carbon-fiber epoxy, fiberglass, metal-matrix composites, ceramic-matrix composites, carbon-carbon, etc. In order to control properties by size-scaling, some microns, sub-micronic, and nanoscopic fillers have been used, thus generating discontinuous composites, micro-composites, and nano-composite materials, respectively. Quantum-composites, the latest in the composite area, have the plasmonic color of coin-metals nanoparticles (e.g., silver, gold) and/or fluorescence, ultra-high/low refractive index, dichroism, etc.; these materials have been deeply investigated by Caseri and the Nicolais research group [25–28, 36].

However, many other ways to combine materials have been conceived and some of these approaches are currently used in the industry (e.g., coatings, multilayers, impregnates, aerogels, etc.). Also, solid solutions (that is, molecular doping of a solid phase) represent a further technological possibility that has great potential and can be used for the same purpose. So far, this approach has been used especially in food packaging [37], polymer electrolytes [38, 39], fluorescent plastics [40, 41], xerographic materials [42], polymer light-emitting diodes [43], drug delivery systems [44], but it can be practically extended to most of the technological fields. These molecular dispersions can involve an organic, inorganic, or organometallic guest (solute) and a variety of solid hosts (solvents). A solid solvent is an amorphous (glassy or rubbery phase) or crystalline medium able to dissolve a solid, liquid, or gaseous solute to generate a solid solution. Since the solvent is the most abundant component of such solid solution, in diluted conditions, the last has mechanical, physical, and chemical properties very close to the solvent. To prevent chemical reactions with the solute, the solvent must necessarily be a chemically inert substance. Polymers are very stable substances, mostly unreactive and not toxic; as a consequence, they can be conveniently used to prepare solid solutions with many organic/inorganic chemical compounds. The use of polymers is widely spread in our lives just for such inert characteristics that they possess. In order to prepare homogeneous systems, amorphous polymers are strictly required; indeed, the solute molecules cannot be "accommodated" in the crystallites contained in the semi-crystalline polymers. Colorless, amorphous polymers (glassy polymers) are frequently used for optical purposes since they are perfectly transparent (absence of light absorption and scattering) and have a refractive index quite close to the glass (i.e., 1.5). These optical-grade polymers are usually named: "optical plastics," and some of them are widely used for fabricating optical components like goggles, optical fibers, color filters, UV limiters, protective windows, etc. The most common optical plastics are the following: polystyrene (PS), poly(methyl methacrylate) (PMMA), polycarbonate (PC), thermoplastic polyurethane (TPU), etc. The use of these polymers to prepare solid solutions with functional solutes constitutes an important technological potentiality, which should be explored in depth to take full advantage of it.

The approach of solids doping (e.g., functional-molecules solubilization in polymers) represents a simple way to convert functionality from a molecular type to a solid-state type. Solid-state functionality has great advantages compared to the molecular one, such as: very simple artefact handling and protection of the functional principle from chemical/thermal degradation since it is incorporated in the polymeric matrix, which has gas-barrier properties for the oxygen and it is thermally an insulator [37]. Also toxicity and biocompatibility of the active "ingredient" become much less critical factors, when the functional group is insulated into a glassy matrix. To avoid releasing the doping molecules outside the polymeric matrix, such molecules should be a nonvolatile species, like a molecule with a high molecular weight or a salt. However, salts hardly dissolve in polymers and solubility represents a very critical issue for these systems.

These special materials can be also designed by looking at nature and trying to mime it (biomimetics [45]). Biomimetic chemistry is addressed to the design and fabrication of new synthetic materials using the strategies adopted by living organisms to produce biological analogs. However, where could start molecular biomimetics in imitating nature? Water could be considered as the most important substance available in nature. Nature has based all living matter on water and it is at the origin of life. Life comes from water and water is the most important solvent available in nature since humans are made of water solutions by 95% and an extremely large number of very important solutions are based on it (the sea, blood plasma, cytoplasm and cytosol of all cells, milk, fruits, etc.). Therefore, it could be very useful to find a solid substance strictly resembling water, a solid with the physical characteristics of water, namely, a "pseudo-water," to be used for simulating water solutions. Polymers are very versatile solids, indeed; they can be rubbery or glassy, electrically conductive or insulating, elastomeric, hydrophilic or hydrophobic, opaque, optically transparent, etc.; as a consequence, in this artificial material class, it is possible to find whatever solid type one is looking for. The possibility to change polymer by changing monomer ("mutans mutandis") is an extraordinary potentiality, an incredible "degree of freedom" to develop new materials since whatever type of molecule could be opportunely functionalized by the introduction of two or more reactive chemical terminals for allowing its polymerization by polycondensation [46]. In the polymer class, a polymeric solid solvent with characteristics similar to water is available and it corresponds to the poly(vinyl alcohol) (PVOH). PVOH is an amorphous solid substance very rich in hydroxyl groups, indeed it is a vinyl polymer with hydroxyl (-OH) as a side group. PVOH appears as artificial solid water, a "polymeric water" we can say, and it possesses "to the touch" all those characteristics of a soft polymer (at room temperature it is in the rubbery state). PVOH has a dipole moment slightly lower than water, however, it is capable of dissolving most solutes that liquid water dissolves (salts, polar molecules, surfactants, drugs, etc.). At ambient conditions, the dielectric constant of pure liquid water is 78.4, while that of solid PVOH is quite close. An important technological potentiality of water is its ability to dissolve organic/inorganic salts and this unique potentiality is mostly kept by PVOH [47]. For example, it is possible to dye PVOH with all common dye salts [48, 49]. Large amounts of cationic and anionic dyes dissolve in PVOH at room temperature, because of its similarity with water. As an example, a PVOH film dyed by methylene blue (0.1% by weight) has been prepared simply by solution casting technology using distilled water as a solvent and it is shown in Fig. 4.2.

As visible, a vitreous (i.e., perfectly transparent) and intensively colored PVOH/methylene blue film can be obtained. However, PVOH can dissolve a variety of salts just like water, indeed many solid ionic conductors, that mimic sea water, and electrolytic capacitors dielectrics are based on it [50–55] and this aspect represents a great potentiality since a large number of colors and fluorescent dyes, radiopaque salts, contrast agents, EMI shields, and other functional molecules are available just

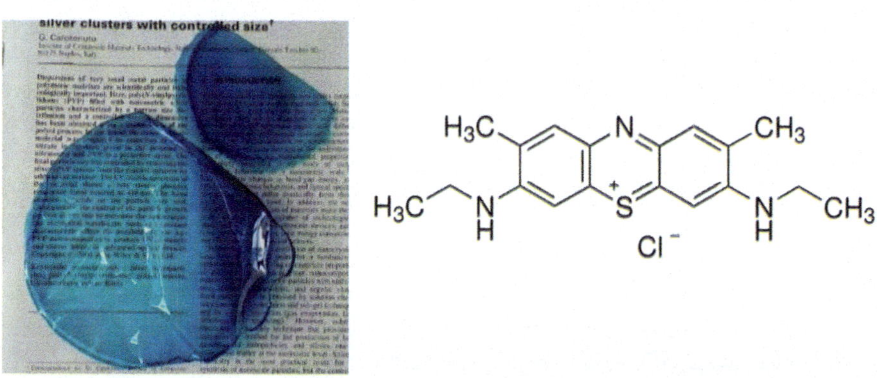

Fig. 4.2 PVOH film dyed by methylene blue (0.1 wt.%) prepared by the solution casting technology

in a salified form. Such PVOH/dye systems can be exploited as color filters, dosimets (in combination with radiochromic centers, since PVOH is a tissue-equivalent material) [55–57], optical recording systems [58–60], colored pastes (slimes), ionizing radiation shields, DNA/PVOH systems [61], and for many other technological applications. Many other types of dyes and organic/inorganic salts (NaCl, $ErCl_3$, $AgNO_3$, etc.) can be dissolved in a PVOH matrix in order to achieve useful functional polymer-based materials [62–66]. A further important point is the change of the PVOH nature from dielectric to electrical conductor as a consequence of the dissociated electrolyte presence. For example, PVOH films doped by $AgNO_3$ have shown a decrease of the direct energy band gap value from 2.30 eV to 1.80 eV by increasing the concentration of $AgNO_3$ dopant from 20 to 80 wt%. This phenomenon is related to the presence of Ag^+ which is a fast-conducting ion that improves the conductivity of the material into which it is incorporated [67, 68]. Consequently, its dissolution within a polymer like PVOH capable of forming a polymer-metal complex is expected to enhance its electrical and optical properties.

However, polar polymers like PVOH could have only limited applications in some technological areas due to their scarce stability to moisture and liquid water. Therefore, solid solutions of nonpolar macromolecules (i.e., hydrophobic or lipophilic solid solutions) are technologically useful and have been developed too.

Obviously, the cited biomimetic approach can be applied also to the solute, in addition to the solid matrix, as described above. Indeed, in nature a very large variety of functional molecules are available [69], these molecules can have functionalities that are very difficult to achieve artificially. The functional molecule, that more than any other has found the most varied technological applications over the years, has been rosin. Rosin or Colophony, also called "Greek pitch" or "violin resin" or "rubber resin," takes its name from the ancient Greek city: Colophon, an Ionian city in Lydia from which it was imported. Nowadays, this substance is mainly imported from southern China and Southeast Asian countries such as Indonesia and Vietnam. Colophony is a natural resin obtained from the solid residue of the distillation of turpentine, which is obtained in turn from the incision of the trunk of pines or other conifers such as firs and larches. Colophony appears as a transparent solid with a color that can vary from light yellow to amber and is mainly made of abietic acid, a diterpene, of molecular formula: $C_{20}H_{30}O_2$, and a quantity of less than 10% by weight of dihydroabietic acid and dehydroacetic acid. Colophony has a density ranging from 1.03 g/cm^3 to 1.12 g/cm^3 and a melting point falling in the interval 60–130 °C; it softens at ca. 70–80 °C. It is insoluble in water, but soluble in many organic solvents such as diethyl ether, chloroform, benzene, and in alkaline aqueous solutions. Since ancient times, this substance has been widely used in various technological sectors including the production of theater masks or sealing wax, as well as for insulating wooden boats. Today the main technological areas for the use of this resin are the following [68–71]:

- In the chemical industry for paints, inks, photocopy and laser-printed paper, sealing wax, soaps, adhesives, pitch, lubricants.
- In the textile industry for the manufacture of non-crumple fabrics.

- In the food industry as a stabilizer for some soft drinks or as a coating for some chewing gum. It is also used in vegetable and emulsified oils in which flavors and aromas are dissolved because it prevents the emulsion from separating from the aqueous fraction over time to form a less dense upper phase.
- In the cosmetic industry, where it is used for the production of depilatory waxes and in the composition of cosmetics.
- In the pharmaceutical industry, where it is used for the coating of microcapsules or nanoparticles, or as an ingredient in ointments and plasters.
- In the diving area as a component of the putty to repair wet suits.

In addition, colophony is also used for electrical insulation and as a deoxidizer in soldering, and also in the optical industry. Further, it can be used to make linoleum, a very resistant and recyclable flooring. There are vast uses of rosin that exploit its frictional force [38], for example:

- In musical instruments, where in particular it is rubbed on the strings of bowed instruments to obtain greater friction and to improve the sound. Or it is applied to the bridges of banjos and banjoleles (always stringed instruments), to prevent the bridge from moving during use.
- In classical dance, in various ballets, and in flamenco, where it is rubbed on the toes and heels as an anti-slip, to increase grip.
- In many sports, where gymnasts, handball players, and rock climbers use it to improve grip, and Olympic throwers rub it on their soles to increase traction on the platform; in baseball, rosin powder is used to optimize control over the ball. In archery, it is used for string maintenance because it increases adhesion and reduces wear and fraying.

Therefore, from the examples given above, it is possible to understand how this chemical compound proves extremely useful in many technological fields of industrial applications.

4.3 The Case Study: Composite Material Based on Rosin in PMMA

Rosin may have an additional extremely important application [72–75]: to provide optical polymers of intensive fluorescence. Indeed, this natural molecule is strongly luminescent [76–78] and, therefore, it can be used to dope optical polymers to generate fluorescent plastics. In addition, the luminescence characteristics of the resulting solid solution strictly depend on the type of host polymer and, therefore, it could be controlled by changing the host. This method to functionalize the polymers is much simpler and cheaper than the currently investigated approaches based on the chemical synthesis. Concerning the mixability of colophony with poly(methyl methacrylate) (PMMA), the colophony has a hydrocarbon skeleton, containing a diene group, and a carboxylic acid group functionalizes it. Similarly, acrylic polymers, like PMMA, contain a hydrocarbon backbone chain and a polar side-group, which consists of an ester function. Consequently, a strong physical interaction between polymer and colophony is possible and such property allows to obtain a solid solution, which is optically transparent. Optical transparency is a relevant characteristic of a fluorescent material since it allows the collection of the emissions of all dissolved molecules.

Among the different optical polymers (i.e., transparent amorphous polymers with refractive index close to 1.5), we selected the amorphous PMMA for the materials development because the PMMA is a very common optical plastic, inexpensive, and has great mechanical performance and very easy industrial processing.

Homogeneous rosin-PMMA solid solutions can be easily obtained by using the solution-casting technology based on chloroform as solvent. The rosin content in PMMA was 1 wt.% for all prepared solid solutions and samples were in the form of thick films.

The optical properties of the produced solid solutions were analyzed by recording the absorption and photoluminescence spectra using a UV-visible spectrophotometer (Perkin-Elmer-Lambda 850) and a Spectrofluorimeter (Perkin-Elmer-LS55), respectively. Both instruments allow the direct investigation of the as-prepared samples in the form of film specimens.

The UV-visible absorption spectrums of rosin dissolved in ethanol is presented in Fig. 4.3a). In the UV region, the sample shows a broad absorption in the 200–300 nm range characterized by a peak centered at 205 nm and a stronger peak at 240 nm that can be referred to as conjugate double bonds present in the rosin structure (see the inset of Fig. 4.3a). The sample is also transparent in the visible range due to the high solubility of rosin in ethanol (absence of light-scattering). Figure 4.3b shows the visible absorption spectrum of the rosin dissolved in solid PMMA. The complete dissolution of rosin in PMMA is confirmed by the low absorbance of light in the visible region.

The luminescence property of the prepared solid solution was thoroughly studied by acquiring the excitation and emission spectra of pure rosin solution in ethyl alcohol and the rosin solid solution in PMMA. Figure 4.4a shows that the strongest emission, centered at 479 nm, originates from the excitation at a wavelength of 400 nm for the pure rosin solution, while, when rosin was dissolved in PMMA,

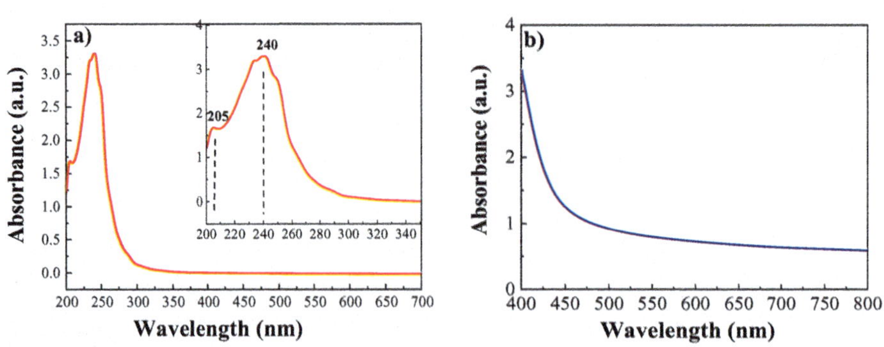

Fig. 4.3 UV-visible absorption spectrum of rosin in ethanol (**a**) and rosin in PMMA (**b**)

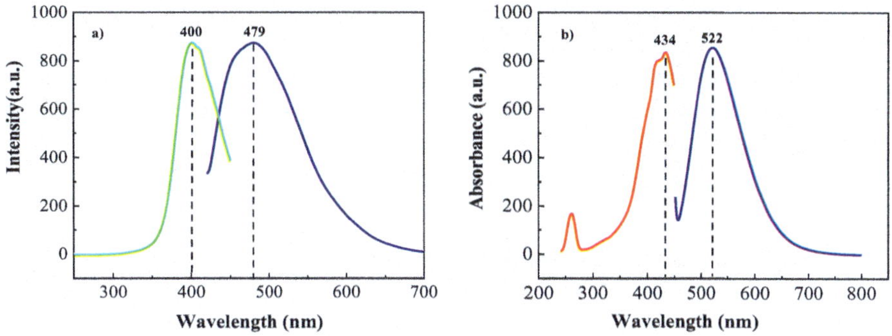

Fig. 4.4 Excitation and emission spectra of rosin dissolved in ethanol (**a**) and rosin dissolved in PMMA (**b**)

Fig. 4.5 Typical image of pure rosin (**a**) and rosin/PMMA amber colored (**c**) and transition to cyan color under UV-light source ($\lambda_{ex} = 365$ nm) excitation (**b**) and (**d**), respectively

excitation and emission redshifts were observed (see Fig. 4.4b). In fact, the strongest emission, centered at 522 nm, was generated by the excitation at a wavelength of 434 nm. This redshift can be ascribed to strong physical interactions between the rosin molecules and the polymeric matrix.

It must be pointed out that, in this spectral region, PMMA does not have luminescence properties. According to spectrofluorimetric results, rosin emits a cyan light under UV excitation (mercury lamp for TLC, 365 nm), while it has an amber color under visible light (see Fig. 4.5a, b). The same amber color and photoluminescence phenomenon has been observed for the rosin solid solution in PMMA under the UV light of a mercury lamp for TLC (365 nm) (see Fig. 4.5c, d). The prepared doped polymer exhibits a stable photoluminescence over extended times since polymer embedding prevents rosin from degradation phenomena.

To the best of our knowledge, the use of these technologically advantageous natural molecules has never been proposed for polymer functionalization in sustainable design applications.

4.4 Nature Power, Empower Nature

> *Ho voglia di volare, di nuotare, d'abbaiare, di muggire, di urlare. Vorrei avere le ali, una corazza, una scorza; vorrei emettere fumo, avere una proboscide, contorcermi, dividermi in ogni luogo, essere in ogni cosa, dissolvermi negli odori, svilupparmi come le piante, scorrere come l'acqua...penetrare ogni atomo, discendere fino al fondo della material—essere materia! (Gustave Flaubert, La tentazione di sant'Antonio, trad. di Agostino Richelmy, Torino, Einaudi, 1990)*

"Life creates conditions conducive to life that's also the design brief for us"— with these words, Janine Benyus in *Biomimicry: Innovation Inspired by Nature* highlights how nature for its extreme creativity in terms of solutions, performance, processes, forms, and sustainability is a model to be imitated [79, 80]. When we choose biomimicry as a design process, we are essentially facing two approaches: [81] Design looking to biology (top-down approach) or Biology influencing design (Bottom-Up approach) (Biomimicry Guild, 2007). The former defines a need or design problem and draws inspiration from solutions used by other organisms or ecosystems to solve it; the latter identifies a natural characteristic or behavior and translates its benefits into design solutions [6]. In addition to these approaches, there are three possible levels of biomimicry to be applied to design problems: the first with a focus on the organism, consisting of imitating the shapes and structures of a particular organism or imitating a part of a whole organism, the second observing the processes and behavior of organisms, and the third imitating complex systems at the ecosystem level. At each level, we can identify five aspects that determine the type of mimicry, and we define design as biomimetic with reference to: appearance (form), material (material), the way it is made (construction), operation (process), and ability (function). The third level, where mimicry involves the entire ecosystem, is considered the most challenging, as it focuses on an aspect that functionally is very difficult to mimic even though it is believed to have high potential in regenerative design on an urban scale by enabling the reconnection and combined evolution of urban and ecological systems. Ecosystem-level biomimicry for regenerative urban design uses the understanding of the patterns and functioning of ecosystems as models for the design of urban spaces and architectures even though the integration of knowledge in both the conception and application of these concepts and methods does not seem to be established yet, and deeper ecological knowledge is needed [82–84], which would allow a better understanding of the functioning of ecosystems and the biophysical structure of the ecosystem and related processes, in order to guide urban design processes with a biomimetic approach.

In his TED talk "What's it like to be a giant sequoia tree?" [85], artist and director Ersin Han Ersin, member of the London-based experiential art collective Marshmallow Laser Feast (MLF), referring to the concept of ecosystem, brings attention back to the philosophical concept of the *umwelt* coined by the biologist Jakob von Uexküll [86], suggesting that living organisms, while sharing the same space, perceive it according to their own personal experience, and, therefore, each species experiences reality in the way its senses perceive it. The speaker then invites the audience to a reflection "What is it like to be a tree?" How does it feel to be at the center of infinite relationships that anchor an entire ecosystem? This question opens the presentation of his design practice that explores the links between the human and natural worlds by bringing them to the audience through immersive sensory experiences. We would like to focus here on the multisensory, mixed-reality installation called "We Live in an Ocean of Air", the work premiered at the Saatchi Gallery in London and winner of the British Animation Award in 2020, in which the umwelt of trees is translated for human understanding, through respiration. In Ersin han Ersin's words: "We take anywhere from 17,000 to 30,000 breaths a day. A third of those breaths coming from the forests and the rest coming from the oceans. This puts us in an intimate relationship with the trees, thousands of times a day. You might think you've never met a giant sequoia before, but in fact, you've been enmeshed and entangled with one every moment you've been alive. In other words, we are as much trees as trees are us." The immersive installation, *We Live in an Ocean of Air* by Marshmallow Laser Feast, incorporates a VR headset to which hand motion trackers, breathing sensors, and heart-rate monitors are attached to allow the audience to discover and interact with the ecosystem surrounding a giant sequoia through breathing. "The experience shows how this giant tree and humans both share similar lung like arborescence in a perfect mirrored complementarity" [10, 87].

What emerges is a desire to encourage a change of perspective in the direction of interdependence, interaction, and collectivity with respect to the human-ecosystem relationship; a view according to which objects and subjects are not unrelated to their interaction, but are constituted in the interaction itself, an intra-action according to which the ability to act "agency" only becomes real in the relationship [88].

It is, therefore, no coincidence that there is currently great interest in the design of artifacts aimed at coworking spaces, "Community of work" [89], introducing positive changes in cognition and perception of emotions, and recognizing that green elements and furniture designed with sustainability in mind not only create an uplifting and relaxing environment but also increase productivity and communication and reduce stress, influencing the subconscious parts of the brain and offering the feeling of a more comfortable environment [90].

Design has always had a social role. With its slogan "places, spaces, people and behaviours" Martino Gamper effectively illustrates the fact that designing means: considering the geographical location and cultural context, the physical space of the artifact, the people who animate these artifacts, and the behavior they will determine, especially in a community and relational dimension [91]. In order to make people feel better in their working spaces, it is, therefore, possible to draw inspiration from the natural world, creating a perfect and harmonious union between nature itself, materials, humans, and technology by designing innovative landscapes inspired by biomimicry and led by a biophilic design approach. Biophilic design [92] is a design strategy that focuses on the relationship between humans and nature and considers the positive effects of nature on people, trying to integrate them into the built environment "Biophilic Design is not about greening our buildings or simply increasing their aesthetic appeal through inserting trees and shrubs. Much more, it is about humanity's place in nature, and the natural world's place in human society, a space where mutuality, respect, and enriching relation can and should exist at all levels and emerge as a norm rather than the exception" [93]. Kellert and Wilson identify six principal biophilic design elements that can be included in an environment to foster the dialogue between mankind and nature: (1) environmental features, such as plants, water, and fire; (2) natural shapes and forms; (3) natural processes; (4) light and space; (5) ground-based relationships; and (6) enhanced human–nature relationships [94]. Biophilic design, by enriching indoor environments with natural elements, improves well-being, happiness, and productivity at work, and according to literature, among biophilic design elements, indoor plants seem to have a high restorative effect on well-being as stress reducers and on cognitive performance increasing the attention capacity [95–99].

Plants are important elements of biophilia and prove to be an excellent model of inspiration.

Plants are endowed with an extremely modern modular, cooperative, and distributed organizational architecture that does not provide centralized control systems but decentralized functions throughout the entire body in a flexibility that makes them adaptable to changing surroundings. The root system is the most relevant part of the plant and operates "as a kind of collective brain or rather intelligence distributed over an area that can be enormous" [100–102].

Plants are one vibrant, hyper-connecteed organism communicating and living through countless biological invisible pathways.

There is a lot of light that we are unable to see that enables and reveals the communication among invertebrates, fish and reptiles, amphibians, birds, fungi, bacteria, and insects [20, 103] but has not yet been found in terrestrial higher plants. That is the invisible biofluorescence phenomenon, and to observe it, we need a substance that absorbs UV radiation and re-emits other radiation with longer wavelengths, which are part of visible light. This condition for which flowers and plants do not exhibit bioluminescence in nature is currently a topic of lively scientific interest and research into modifying it [104, 105].

As previously described, rosin, a yellow, solid, transparent plant resin that is a residue from the distillation of turpentines (resins of conifers Pinus palustris and other pines of the Pinaceae family), has been investigated because of its unique luminescence. Because of this special characteristic and because a strong physical interaction is possible between polymer and rosin, leading to the formation of a solid and optically transparent solution, this natural molecule appears perfectly suited to versatile and sustainable design projects.

Table a: Rosin concept

Interplay is a colorful installation, that makes us aware of the wonderful achievements that people can reach by acting collectively and caring for each other and the whole ecosystem. *Interplay* is a transparent surface in PMMA (an optical plastic, inexpensive with great mechanical performance and very easy processability), located within shared workspaces designed according to a biophilic approach. It is an always-on optical sensor that encourages a collective creative environment. When more than one person approaches the surface, the wall lights up and the atmosphere in the room becomes more colorful, the more people interact with it. An auditory soundscape, which captures ambient sounds and in real time and mixes them with a natural soundscape (aligned with the outdoor natural environment), completes the whole experience.

Interplay enlights the creative energy that two or more people are able to release when they are together.

Table b: Interplay concept

However, it is not only about a relationship between humans, *Interplay* also explores the relationship with nature and the surrounding environment by monitoring plants and for "the magic" to happen, the whole ecosystem must be in harmony. Once there is no motion in the room, the surface rests with a low illumination (ready to be stimulated) but if the plants in the space are not in good condition, the system will not be activated.

How does it work? The system is based on the rosin's ability to "sparkle" when under UV light and two information/interaction modes are activated by sensors monitoring the surroundings. The surface is illuminated thanks to a number of UV lamps (with low energy consumption) which, depending on different inputs, are activated or not.

Interaction with plants: communication with plants is structured to monitor their needs with sensors such as temperature, phototransistors, or humidity sensors, all managed via an app that receives information from the sensors and in real-time displays their status by switching on a number of UV lamps that activate the composite material based on rosin (when the plants are in good condition the surface lights up). A healthy environment, a favorable condition, will also provide a visually relaxing environment.

Interaction with people: Interaction with people is structured by sensor technologies that detect the presence of people in the proximity of the surface, gradually activating UV lamps and soundscape microphones and loudspeakers (the presence of several people interacting results in an increasing creative energy).

Interplay is based on several works whose findings suggest that workplace productivity increases due to the impact of a biophilic visual design on cognitive performance, as well as satisfaction and sense of psychological support of the workplace. These works indicate that immersive interventions, with a multi-sensory approach, including sounds and images that recall nature and incorporate indoor plants into the design, reduce stress and improve: i) satisfaction with the indoor environment, ii) perceived productivity, and iii) workplace conditions [107]. *Interplay* unleashes the invisible energy of a relational space and empathizes with its inhabitants, giving to the magic of the relationship an unexpected allure.

Innovation in material design opens up surprising new perspectives in which surfaces become vibrant interfaces that interact with the world in a more sustainable approach, respectful of, and inspired by, the environment, generating atmospheres of well-being that borrow new lifeblood from nature.

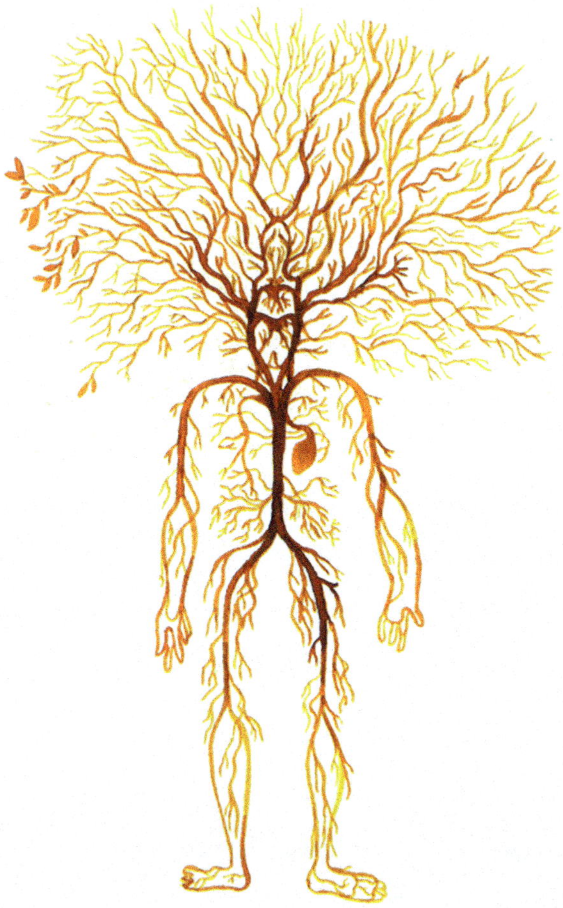

Table c: Biomimicry inspiration concept

Note: Concept tables are original watercolor illustrations realized for the authors by Vincenzo Del Vecchio.

References

1. Yeang K (1974) Bionics: the use of biological analogies for design. Archit Assoc Q VI(2):48–57
2. Salvia G, Rognoli V, Levi M (2009) Il progetto della natura. Gli strumenti di biomimesi per il design. Franco Angeli, Milano
3. Langella C, Santulli C (2017) Processi di crescita biologica e Design parametrico. MD J 3:14–27
4. Hooker G, Smith E (2016) AskNature and the biomimicry taxonomy. Insight 19(1):46–49
5. Bar-Cohen Y (2005) Biomimetics: biologically inspired technologies, CRC Press, Boca Raton
6. Katiyar NK, Goel G, Hawi S et al (2021) Nature-inspired materials: emerging trends and prospects. NPG Asia Mater 13(56)
7. Langella, C. (2020). Complicità bioniche. In BIONICA E DESIGN. Carmelo Di Bartolo e il Centro Ricerche IED esperienze memorabili da 30 protagonisti 215–232. Blucher
8. Svensson E (2017) Material characterization of 3D-printed energy-absorbent polymers inspired by nature. Master's thesis, Chalmers University of Technology, Göteborg, Sweden
9. Darmanin T, Guittard F (2015) Superhydrophobic and superoleophobic properties in nature. Mater Today 18:273–285
10. Gust D, Moore TA, Moore AL (2009) Solar fuels via artificial photosynthesis. Acc Chem Res 42:1890–1898
11. Mastinu A, Kumar A, Maccarinelli G, Bonini SA, Premoli M, Aria F, Gianoncelli A, Memo M (2019) Zeolite Clinoptilolite: therapeutic virtues of an ancient mineral. Molecules 24:1517
12. Ivanova II, Knyazeva EE (2013) Micro–mesoporous materials obtained by zeolite recrystallization: synthesis, characterization and catalytic applications. Chem Soc Rev 42(9):3671–3688
13. Kowalczyk P, Sprynskyy M, Terzyk AP, Lebedynets M, Namieśnik J, Buszewski B (2006) Porous structure of natural and modified clinoptilolites. J Colloid Interface Sci 297:77
14. Schiavo L, Boccia V, Aversa L, Verucchi R, Carotenuto G, Valente T (2023) Natural clinoptilolite nanoplatelets production by a friction-based technology. Mater Proc 14(1):11
15. Harper M (2000) Sorbent trapping of volatile organic compounds from air. J Chromatogr A 885(1–2):129–151
16. Diboune M, Nouali H, Soulard M, Patarin J, Rioland G, Faye D, Daou TJ (2020) Efficient removal of volatile organic compounds by FAU-type zeolite coatings. Molecules 25(15):3336
17. Djaeni M, Kurniasari L, Sasongko SB (2015) Preparation of natural zeolite for air dehumidification in food drying. Int J Sci Eng 8(2):80–83
18. Hernández MÁ, Hernández GI, Portillo RI, Velasco MDLÁ, Santamaría-Juárez JD, Rubio E, Petranovskii V (2023) Influence of chemical pretreatment on the adsorption of N2 and O2 in Ca-clinoptilolite. Separations 10:130
19. Inglezakis VJ (2005) The concept of "capacity" in zeolite ion-exchange systems. J Colloid Interface Sci 281(1):68–79
20. Olson ER, Carlson MR, Ramanujam VMS et al (2021) Vivid biofluorescence discovered in the nocturnal springhare (Pedetidae). Sci Rep 11:4125
21. Wang Q, Gossweiler G, Craig S et al (2014) Cephalopod-inspired design of electro-mechano-chemically responsive elastomers for on-demand fluorescent patterning. Nat Commun 5:4899
22. Su YH, Tu SL, Tseng SW, Chang YC, Chang SH, Zhang WM (2010) Influence of surface plasmon resonance on the emission intermittency of photoluminescence from gold nano-sea-urchins. Nanoscale 2(12):2639–2646

23. Grandi GM (2020) Adamo è morto. Cy-born life vs Bio-aptic environments. In: BIONICA E DESIGN. Carmelo Di Bartolo e il Centro Ricerche IED esperienze memorabili da 30 protagonisti 320. Blucher, pp 310–322
24. Gandhi MV, Thompson BD (1992) Smart materials and structures. Springer
25. Carotenuto G, De Nicola S, Ausanio G, Massarotti D, Nicolais L, Pepe GP (2014) Nanoscale Res Lett 9:475. https://doi.org/10.1186/1556-276X-9-475
26. Nicolais L, Carotenuto G (eds) (2005) Metal-polymer nanocomposites. Wiley, Hoboken
27. Nicolais L, Carotenuto G (eds) (2013) Nanocomposites: in situ synthesis of polymer-embedded nanostructures. Wiley, Hoboken
28. Nicolais L, Carotenuto G (2015) Functional and multifunctional nanostructured materials. In: Breuer KFU (ed) Multifunctionality of polymer composites—challengs and new solutions. Elsevier, pp 842–857
29. Carotenuto G, Nicolais L (2012) Graphene-polymer composites. In: Nicolais L, Borzacchiello A (eds) Wiley encyclopedia of composites, II edition. Wiley, Hoboken
30. Carotenuto G, Giannini C, Siliqui D, Nicolais L (2011) Preparation of polymer-embedded metal and metal sulfide clusters. In: Wiley encyclopedia of composites, pp 1–4
31. Peluso A, Pagliarulo V, Carotenuto G, Pepe GP, Davino D, Visone C, Longo A, Capezzuto F (2009) Microw Opt Technol Lett 51(11):2774–2777
32. Carotenuto G, Longo A, De Nicola S, Nicolais F Advanced thermochromic plastics for optical temperature sensing. In: Somani PR (ed) Chromic materials, phenomena and their technological applications. Applied Science Innovation
33. Carotenuto G, Longo* A, Hison CL (2009) Tuned linear optical properties of gold-polymer nanocomposites. J Mater Chem 19(32):5744–5750. Longo A, Pepe GP, Carotenuto G, Ruotolo A, De Nicola S, Belotelov VI, Zvezdin AK (2007) Optical emission studies in Au/Ag nanoparticles. Nanotechnology 18(36):365701–365706
34. Nicolais F, Carotenuto G (2013) Corporate Security Innovazione per la tutela del Made in Italy. In: Design intersections—il pensiero progettuale itermedio, a cura di Patrizia Ranzo e Carla Langella. FrancoAngeli S.r.l., Milano, Italy, pp 123–133. ISBN: 882041144X
35. Caseri W (2000) Nanocomposites of polymers and metals or semiconductors: historical background and optical properties. Macromol Rapid Commun 21(11):705–722
36. Stoleru E, Brebu M (2021) Stabilization technique of essential oils by incorporation into biodegradable polymeric materials for food packaging. Molecules 26:6307
37. Harvey PE (1989) Polymer electrolytes. J Power Sources 26(1–2):23–32
38. Lusi M (2018) A rough guide to molecular solid solutions: design, synthesis and characterization of mixed crystals. CrystEngComm 20:7042–7052
39. Peterson KA, Zimmt MB, Fayer MD et al (1989) Fluorescence depolarization of chromophores in polymeric solids. Macromolecules 22:874–879
40. Calderon OG, Guerra JM, Costela A, Garcia-Moreno I, Sastre R (1997) Laser emission of a flash-lamp pumped Rhodamine 6 G solid copolymer solution. Appl Phys Lett 70:25–27
41. Loutfy RO, Ong BS (1984) Spectroscopic studied of the compatibility of small molecules in polymers. Can J Chem 62:2546
42. Liu J, Shi Y, Yang Y (2001) Improving the performance of polymer light-emitting diodes using polymer solid solutions. Appl Phys Lett 79:578–580
43. Budiman A, Nurani NV, Laelasari E, Muchtaridi M, Sriwidodo S, Aulifa DL (2023) Effect of drug-polymer interaction in amorphous solid dispersion on physical stability and dissolution of drugs: the case of alpha-mangostin. Polymers 15(4):3034
44. Vincent JFV (2009) Biomimetics—a review. Proc Inst Mech Eng H J Eng Med 223(8):919–939
45. Calvert PD (1994) Polymers for new materials. Polymer 35:4484–4488
46. Ghoshal D, Bhattacharya D, Mondal D, Das S, Bose N, Basu M (2019) Methylene blue/PVA composite film for flexible, wide-scan UV-vis laser cut-off filter. Mater Res Exp 6:075332
47. Sugita H, Ebata S, Suzuki H, Kurg S, Yoneda E, Yoda K, Mashima H, Fukuma S, Kanai D, Yamamura T, Manabe Y, Tateshima K (2023) Kolor-filter production. Displays 77:102393

48. Zaitsev BE, Dzhuraey RR, Kotov AV (1993) PVA-based light filter for UV and visible spectra. J Appl Spectrosc 59:612–613
49. Fadil F, Adli FA, Affandi NDN, Harum AM, Alam MK (2020) Dope-dyeing of polyvinyl alcohol (PVA) nanofibers with remazol yellow FG. Polymer 12(12):3043
50. Wong JIC, Ramesh S, Jun HK, Liew CW (2021) Development of poly(vinyl alcohol) (PVA)-based sodium ion conductors for electric double-layer capacitor application. Mater Sci Eng B 263:114804
51. Kanbara T et al (1989) New lithium salt ionic conductor using poly(vinyl alcohol) matrix. Chem Lett 11:1913–1916
52. Kanbara T, Inami M, Yamamoto T (1991) New solidstate electric double-layer capacitor using poly(vinyl alcohol)-based polymer solid electrolyte. J Power Sources 36:87–93
53. Rajendran S, Sivakumar M, Subadevi R (2003) Effect of salt concentration in poly(vinyl alcohol)-based solid polymer electrolytes. J Power Sources 124:225–230
54. Kim YD, Jo YK, Jo NJ (2012) Electrochemical performance of poly(vinyl alcohol)-based solid polymer electrolyte for lithium polymer batteries. J Nanosci Nanotechnol 12:3529–3533
55. Anh VT, Nghiep TD, Giap TV (2018) To determine dose response curves of dyed polyvinyl alcohol films irradiated with gamma-rays. Open J Saf Sci Technol 8:13–19
56. Akhtar S, Hussain T, Shahzad A, Islam Q, Hussain MY, Akhtar N (2013) Radiation induced decoloration of reactive dye in PVA films for film dosimetry. J Basic Appl Sci 9:416–419
57. Akhtar S, Hussain T, Shahzad A, Islam Q (2013) The feasibility of reactive dye in PVA films as high dosimeter. J Basic Appl Sci 9:420–423
58. Ushamani M, Sreekumar K, Kartha CS, Joseph R (2004) Methylene blue doped polymers: efficient media for optical recording. J Mod Opt 51:743–755
59. Solano C, Martinez-Ponce G, Castaneda C (2006) Dye-polyvinyl alcohol films: molecular weight and hydrolysis degree influence on optical recording. Appl Opt 45(21):5207–5211
60. Hanczyc P, Norden B, Akerman B (2011) DNA in a polyvinyl alcohol matrix and interactions with three intercalating cyanine dyes. J Phys Chem B 115(42):12192–12201
61. Ahmed AAA, Al-Hussam AM, Abdulwahab AM, Ahmed ANAA (2018) The impact of sodium chloride as dopant on optical and electrical properties of polyvinylalcohol. AIMS Mater Sci 5(3):533–542
62. Hamdalla TA, Hanafy TA, Bekheet AE (2015) Influence of erbium ions on the optical and structural properties of polyvinyl alcohol. J Spectrosc 204867
63. Kumar GV (2017) Enhancement of electrical conductivity in polyvinyl alcohol films by doping with Nd3+ ions. Int J Res-Granthaalayah 5(4):87–91
64. Jabbar WA, Habubi NF, Chiad SS (2010) Optical characterization of silver doped poly(vinyl alcohol) films. J Ark Acad Sci 64(21)
65. Bhosale S (2023) Opto-electrical characteristics of pure and silver nitrate-doped polyvinyl alcohol films. Mater Today Proc 80(2):1348–1352
66. Wrigley S (ed) (2011) Functional molecules from natural sources. Royal Society of Chemistry
67. Wilbon PA, Chu F, Tang C (2013) Progress in renewable polymers from natural terpenes, terpenoids, and rosin. Macromol Rapid Commun 34(1):8–37
68. Karlberg AT, Hagvall L (2020) Colophony: rosin in unmodified and modified form. In: Kanerva's occupational dermatology. Springer, Berlin, pp 607–624
69. Yadav BK, Gidwani B, Vyas A (2016) Rosin: recent advances and potential applications in novel drug delivery system. J Bioact Compat Polym 31(2):111–126
70. Silvestre AJ, Gandini A (2008) Rosin: major sources, properties and applications. In: Monomers, polymers and composites from renewable resources. Elsevier, pp 67–88
71. Smith JH, Woodhouse J (2000) The tribology of rosin. J Mech Phys Solids 48(8):1633–1681
72. Maiti S, Ray SS, Kundu AK (1989) Rosin: a renewable resource for polymers and polymer chemicals. Prog Polym Sci 14(3):297–338
73. Colín-Gómez M et al (2022) Synthesis and evaluation of new coatings based on chemically modified rosin and PMMA for steel substrates. Mater Lett 318:132224

74. Kugler S et al (2019) Advances in rosin-based chemicals: the latest recipes, applications and future trends. Molecules 24(9):1651
75. Hong GAO, Liwei DOU (2023) Research progress of Rosin-based fluorescence derivatives. Biomass ChemEng 57(2)
76. Ge M et al (2021) Seeking brightness from nature: sustainable carbon dots-based AIEgens with tunable emission wavelength from natural rosin. Chem Eng J 413:127457
77. Cai X-M et al (2021) BioAIEgens derived from rosin: how does molecular motion affect their photophysical processes in solid state? Nat Commun 12(1):1773
78. Benyus JM (2002) Biomimicry: innovation inspired by nature, 1st edn. Harper Perennia, New York
79. Langella C (2007) Hybrid design. Progettare tra tecnologia e natura, FrancoAngeli
80. Aziz MS, El Sherif AY (2016) Biomimicry as an approach for bio-inspired structure with the aid of computation. Alex Eng J 55(1):707–714
81. Blanco E, Pedersen Zari M, Raskin K, Clergeau P (2021) Urban ecosystem-level biomimicry and regenerative design: linking ecosystem functioning and urban built environments. Sustain For 13:404
82. Hayes S, Desha C, Gibbs M (2019) Findings of case-study analysis: system-level biomimicry in built-environment design. Biomimetics 4:73
83. Pedersen Zari M (2018) Regenerative urban design and ecosystem biomimicry. Routledge, London
84. https://www.ted.com/talks/ersin_han_ersin_what_s_it_like_to_be_a_giant_sequoia_tree
85. von Uexküll J, Kriszat G (1967) Ambiente e comportamento, tr. it. di P. Manfredi, Il Saggiatore, Milano
86. Marshmallow laser feast—focus #3: we live in an ocean of air. Fabbula 2018. https://fabbula.com/artists/we-live-in-an-ocean-of-air-by-marshmallow-laser-feast/
87. Barad K (2007) Meeting the universe halfway: quantum physics and the entanglement of matter and meaning. Duke University Press, Durham
88. Rus A, Orel M (2015) Coworking: a community of work. Teorija in Praksa 52:1017–1038
89. Nieuwenhuis M, Knight C, Postmes T, Haslam SA (2014) The relative benefits of green versus lean office space: three field experiments. J Exp Psychol Appl 20(3):199–214
90. Quinz E (2020) Contro l'oggetto. Conversazioni sul design, Quodlibet, Habitat. ISBN 9788822903266
91. Kellert S, Calabrese E (2015) The practice of biophilic design
92. Kellert S, Heerwagen J (2008) Preface. In: Kellert SR, Heerwagen J, Mador P (eds) Biophilic design. Wiley, Hoboken
93. Kellert SR, Wilson EO (2008) Biophilia. Hum Ecol:462–466
94. Gillis K, Gatersleben B (2015) A review of psychological literature on the health and wellbeing benefits of biophilic design. Buildings 5:948–963
95. Fadda R, Congiu S, Roeyers H, Skoler T (2023) Elements of biophilic design increase visual attention in preschoolers. Buildings 13:1160
96. Dijkstra K, Pieterse ME, Pruyn A (2008) Stress-reducing effects of indoor plants in the built health care environment: the mediating role of perceived attractiveness. Prev Med 47:279–283
97. Kaplan R, Kaplan S (1989) The experience of nature: a psychological perspective. Cambridge University Press, Cambridge
98. Raanaas RK, Evensen KH, Rich D, Sjøstrøm G, Patil G (2011) Benefits of indoor plants on attention capacity in an office setting. J Environ Psychol 31:99–105
99. Mancuso S (2017) Plant revolution. Ed. Giunti, Firenze
100. Ciszak M, Comparini D, Mazzolai B, Baluska F, Arecchi FT, Vicsek T et al (2012) Correction: swarming behavior in plant roots. PLoS One 7(3)
101. Baluška F, Mancuso S (2007) Plant neurobiology as a paradigm shift not only in the plant sciences. Plant Signal Behav 2(4):205–207. https://doi.org/10.4161/psb.2.4.4550
102. Daney de Marcillac W, Nguyen LTP, Aracheloff C, Berthier S, Schöllhorn B (2021) Bright green fluorescence of Asian paper wasp nests. J R Soc Interface 18:20210418

103. Li B, Chen R, Zhu C, Kong F (2021) Glowing plants can light up the night sky? A review. Biotechnol Bioeng 118(10):3706–3715
104. Mitiouchkina T, Mishin AS, Somermeyer LG et al (2020) Plants with genetically encoded autoluminescence. Nat Biotechnol 38:944–946
105. Aristizabal S, Byun K, Porter P, Clements N, Campanella C, Li L, Bauer B (2021) Biophilic office design: exploring the impact of a multisensory approach on human well-being. J Environ Psychol 77:101682

Francesca Nicolais holds a PhD in Communication Sciences and is a researcher at the University Suor Orsola Benincasa. Her research interests focus on design, materials and Digital Humanities. She coordinates experiential projects for Lapej, a production company, and is co-curator of Riot studio, a space dedicated to creativity, cultural research, and innovation. She has been the co-organizer of TEDxNapoli since 2012 and co-inventor of a European patent on the preparation of fluorescent nanoparticles for corporate security.

Angela Longo holds a PhD in Materials and Structures Engineering with a degree in Chemical Engineering and is currently a researcher at the Institute for Polymers, Composites and Biomaterials—National Research Council (IPCB—CNR) of Naples. Her current scientific interests are related to the development of new methods to prepare nanostructured materials as polymer-embedded inorganic nanostructures. Furthermore, her interests include the design and development of advanced devices for electronic, optoelectronic, and biomedical application fields. Several dozen of her scientific papers have appeared in international journals.

Mariano Palomba is a researcher at IPCB-CNR. He holds a degree in Chemical Engineering and a PhD in Materials and Structural Engineering. His current research interests focus on functional nanocomposite materials, metal-polymer nanocomposites, polymer nanostructures, nanocoatings of graphite supported on polymer, new electrical conductors, chemical sensors, chemical-physical characterization of these materials, with particular reference to microscopic techniques of morphological characterization (SEM and TEM), and calorimetric techniques (TGA, DSC and DMA).

Loredana Schiavo received her degree in industrial chemistry at the University of Naples "Federico II" and her PhD in Physics at the University of Trento, Italy. She is a researcher at IPCB-CNR. Her scientific research activity is mainly focused on the development of functional nanomaterials and nanocomposites for energy applications. She has a strong research interest in nature-made materials from an application-oriented point of view.

Gianfranco Carotenuto is a materials science scholar involved in the investigation of nature-made materials from an application-oriented viewpoint. His most recent interest concerns the development of green electrical devices (e.g., water sensors, thermistors, ionic conductors) based on mineral zeolites (clinoptilolite). An interdisciplinary research approach, lying at the interface among chemistry, physics, and engineering, has been applied to natural materials widely distributed in the world, such as diatomite, clinoptilolite, colophony, etc.

Chapter 5
Crafting Ecosystems: Algae, Echinoids, and Algorithms for a New Biomimetic Design Experience

Francesco Dell'Aglio, Enza Migliore, and Chiara Scarpitti

Abstract In 2020, during the COVID-19 pandemic, the information system became an overwhelming and uncontrolled phenomenon. In response, between 2020 and 2022, *Crafting Ecosystems* developed a hybrid design system to highlight the crucial role of design in the information age, fostering the importance of preserving natural ecosystems. The project is focused on a dual temporal perspective, which was evident during subsequent pandemic lockdowns: the speed of media information through continuous digital flows and the slowness of natural ecosystems with their inexorable evolutionary times.

Crafting Ecosystems is a design-driven system that blends the infosphere with nature, critical thinking, and empathy to encourage awareness of how planetary emergencies depend on the interconnected balance between all living and nonliving things, including animals, plants, and information. The project demonstrates the impact of digital information flows by facilitating real-time communication between natural and artificial ecosystems through shaping a seaweed and a sea urchin as biological systems informed by digital data flows.

Keywords Practice-based research · Biodesign · Biomimetic · Data visualization · Speculative ecologies · Sustainable ecosystems · Visual design · Algorithm design

F. Dell'Aglio
Università degli Studi di Napoli Federico II, Naples, Italy

E. Migliore
SUSTech School of Design, Shenzhen, China

C. Scarpitti (✉)
University of Campania "Luigi Vanvitelli", Aversa, Italy
e-mail: chiara.scarpitti@unicampania.it

A. J. V. de Arruda, F. L. Palombini (eds.), *Biology, Biomimetics and Natural Design*, Environmental Footprints and Eco-design of Products and Processes,
https://doi.org/10.1007/978-3-031-65877-8_5

5.1 Ecological Perspectives and Transdisciplinary Design: A Necessary Dialogue

In its complexity, design culture is increasingly interconnected with other disciplines and interacting with knowledge beyond its research field. The very role of design is evolving in response to current emergencies due to its ability to intercept critical issues and unveil forthcoming ones. The interdependence between heterogeneous knowledge is essential in understanding the present condition and outlining possible scenarios as trajectories of critical reflection and collective practice.

This essay, characterized by a strong experimental and practice-based nature, is inherently transdisciplinary, hybridizing fields such as algorithmic design, data visualization, natural sciences, visual graphics, and contemporary philosophy.

The proposed contribution illustrates the *Crafting Ecosystems* project that started in 2020, prompted by online dialogues and brainstorming between Italy and China during the alienating pandemic seen from the perspective of media information. During the COVID-19 emergency, the information system, governed by data flows, emerged as one of the most overwhelming and uncontrolled phenomena. In this scenario, which lasted until 2022, *Crafting Ecosystems* has built and consolidated an interactive design system to shed light on preserving natural ecosystems and urgently define the substantial role of design in the information age.

Concerning technological innovation, the COVID-19 pandemic has profoundly altered the relationships between humans, technology, and network data flows through forced acceleration, radically reshaping their dynamics. Imposed smart working, abuse of social networks, and web-filtered information have become common phenomena of a new way of experiencing social relations and the system that structures our daily lives.

The emergency has challenged the global system of interactions and definitively shifted the center of its focus from humans to the planet. We must take an ecologically conscious approach to create a truly sustainable design project. That means reevaluating our strategies, tools, and overall approach to protect the environment. The crisis marked by the pandemic is a tangible manifestation of the decline of our human civilization, but it also shows another face of the world, hitherto unknown. Since the philosophical consideration of Slavoj Zizek is that every crisis is an opportunity, [1] the central question now concerns how to rebuild our social and political systems in a way that respects ecological limits and works for all species, human and nonhuman. This approach involves looking at the planetary crisis produced by humans, defined as the Anthropocene, as an opportunity to reverse course and look at the world in a pluriverse and nonextractive way [2].

To quote the philosopher Timothy Morton, his viewpoint looks at humans as a geophysical force acting on a planetary scale. Ecological consciousness is awareness of unexpected relationships, such as the acceptance that humans are made of nonhuman identities, like bacteria, for instance, with which we live in symbiosis. "It means becoming sensitive, receptive, and nonhuman, within and without" [3].

Ecological thinking is a virus that infects many areas of thought (...) "It has to do with love, loss, despair, and compassion. It has to do with capitalism and what might exist after capitalism. (...) It has to do with doubt and confusion. It has to do with the concepts of space and time. (...) It has to do with society. It has to do with coexistence" [4].

In contiguity with these reflections regarding an ecosystemic perspective, the project's orientation expresses an anti-hierarchical conception among human-nature entities, according to which humans are no longer the planet's focus but all integrated and interacting species. This point of view implies a universalistic and total integration, thinking of humans with ecosystems, promoting coexistence among different species, living and nonliving entities.

From this perspective, the project's orientation toward principles of biomimicry and critical ecology does not only act at the level of visual storytelling, valuing sustainability principles. Instead, it intervenes by raising awareness of the problem, working directly on data translation to direct observer interaction, and determining real impacts, as we will see in the project description.

The contribution starts with biomimetic exploration by selecting two living organisms from the marine realms, a seaweed and a sea urchin. The goal is to open a performative path of actions and everyday practices capable of creating new systems of coexistence between us and the natural world surrounding us.

Given the unsustainability of our current world, it is through the activity of individual people involved in concrete situations that a new look at the ordinary can emerge, aligning with the magnificence of nature and the nonhuman. The current design culture needs the overarching narratives of modernity to guide it. As we face the challenges of living in a dark age, we must question ourselves and consider the possibility of new ways of thinking and acting responsibly. This calls for a different approach to design [5] (Figs. 5.1 and 5.2).

This practice-based research project, therefore, acts in the present through practices of metaphorical data coding and translation, a kind of rewriting of the signs detected from the two selected organisms. The design idea elaborates new dialogic universes by using and blending multiple dimensions of design: the parametric and cognitive, the experiential and laboratory, and the craft and technological.

In the graphs above, the boards combine artisanal silkscreens and meta-scientific languages, creating a new mode of visual communication, overlayed and mixed media. It is a hybrid narrative expedient, suitable for making visible the different media used, digital and otherwise, through a transition between different types of papers and prints. The visual project does not intend to reveal the system's functionality in a didascalic way but, on the contrary, to deepen it by creating artifacts that can induce a more sophisticated and "contact with the problem" experience [6].

The silkscreens created express physical transitions that narrate new forms of proximity and intermingling between humans' digital data and nature. The aim is to reconnect tiny and invisible forms of life—in this case, organisms and plants—within a broader digital flow's perception to understand their implicit interdependencies.

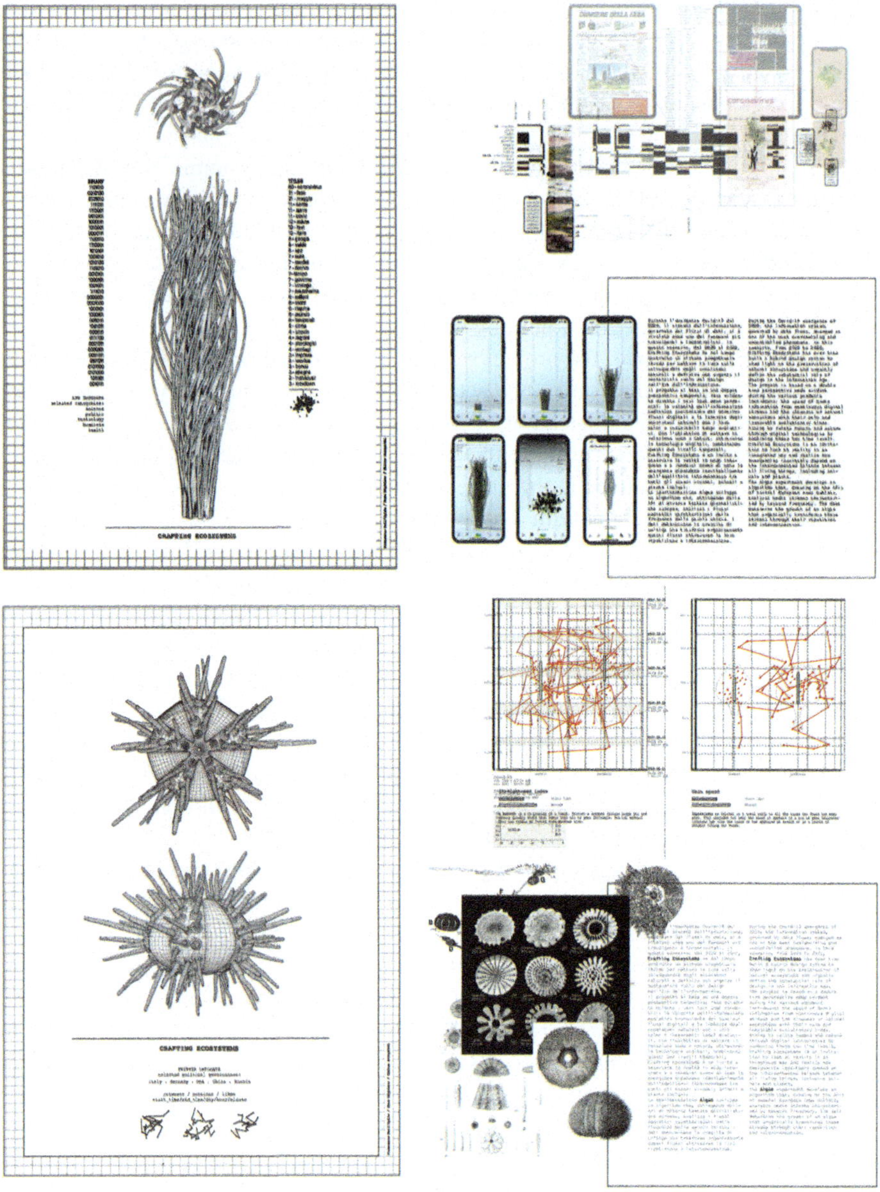

Fig. 5.1 The illustrations represent the study of the layout of the final two explanatory boards of Crafting Ecosystems, related to seaweed and sea urchins. Among mixed media: black and white silkscreen print in numbered series, acetate paper sheets representing scientific study graphs, opal paper sheets representing collage and operation of the app, and conceptual abstract

Fig. 5.2 The illustrations represent the study of the layout of the final two explanatory boards of Crafting Ecosystems, related to seaweed and sea urchins. Among mixed media: black and white silkscreen print in numbered series, acetate paper sheets representing scientific study graphs, opal paper sheets representing collage and operation of the app, and conceptual abstract

By highlighting desired ecological scenarios, the project becomes a strategic vector for dialogic processes where the narrative is also as significant as the result: the goal is to open up questions and to understand what role design might be called upon to play in the immediate future, acting first and foremost from the perspective of change [7].

Referring to the visual communication manifesto drafted in 1964 by Ken Garland, First Things First, meaning priorities, the concept of emergency should be reexamined, defining this condition as a permanent and implicit feature of contemporary society. In this view, design takes on political and educational significance by raising collective consciousness and identifying established critical issues and embryonic emergencies toward a shared sense of earthly citizenship [8].

In the emergent state, paradoxically, the positivistic view that the designer is a problem-solver vanishes altogether, asserting itself instead as one who identifies fragilities, remaining permanently in an inquisitorial state, lacking absolute or permanent solutions, but creating active short circuits, triggers to change. "Compromised, torn, interrupted Nature is, in fact, not only that of the plant world but also that balance between all living species, that coexistence between vital worlds that our anthropocentric culture has progressively ignored. (...) we find ourselves today in full, with our responsibilities as dominant species and our culture as colonizers of ourselves" [9].

So, the design culture is also transformed through changing communication flows as a pioneer of desired narratives and scenarios. In this sense, critical projects such as Crafting Ecosystems, understood as devices of divergent thinking, now assume the capacity to act in reality through their narrative existence by operating an awareness of new behaviors and human-world interactions (Fig. 5.3).

From a transdisciplinary perspective, the research project proposed here draws on different disciplines, from natural sciences to parametric design, from material experiments to 3D modeling. This combination of practices is intended to mediate the relationship between humans and organisms at various scales, ranging from the intangibility of environmental data to the scale of the visible, from the microscopic to the ecosystemic. These are aimed to highlight ways of feeling, perceiving, and experiencing the nature to which humankind belongs. In this way, they become an invitation to consider an alternative knowledge of the world [10].

The materials and media that *Crafting Ecosystems* adopts are closely dependent on the transdisciplinarity between heterogeneous knowledge, skills, and theoretical frameworks, and they change over time, moving from the solid state to the gaseous state to hybridization with digital streams. The dematerialization of design through

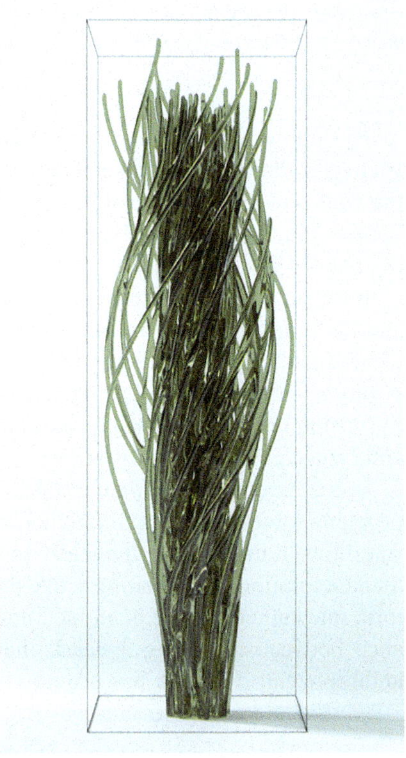

Fig. 5.3 The photograph represents one of two illustrative frameworks of *Crafting Ecosystems*, made through different techniques and media for its scientific and artistic representation. Materials: glass, bleached wood. Dimensions: 100 × 70 cm

the digital contributes to a new materiality. Transdisciplinary tools and expressive design modalities—such as installations, videos, infographic maps, material and process experiments, and technological apps—create a "sci-fi" and hybrid landscape that is very intriguing and multi-dimensional. Hybrid and post-digital, the resulting materiality is synesthetic, vibrant, and multi-dimensional.

Simultaneously adopting the operational protocols of scientific knowledge and the artistic approaches proper to humanistic ones, the project reinterprets reality, enacting migrations of thought that transcend traditional sectorial dichotomies for an increasingly interconnected system of society, nature, and technology.

Let's think about the problems of Climate Change and what the COVID-19 pandemic has clarified. Due to the impact of human industrialization, the relentless alteration of the ecosystem is leading humans to an increasingly widespread awareness that the planet's resources are not infinite, even to the point of a different interaction with them.

It is about designing and acting from nature and the logic of intermingling, human and nonhuman, and experiencing life as a concrete path, emancipated from the prevailing market logic, in the will to subvert a system that is harmful to both humans and the ecosystems to which they belong.

5.2 The Post-anthropocentric Vision of Crafting Ecosystems Through Algae and Echinoid

Crafting Ecosystems is an invitation to look at reality in an integrated way and realize how emergencies inevitably depend on the interconnected balance between all living things, including animals and plants. Through a multidimensional narrative between 3D models, algorithms, videos, screen prints, and mixed media, the project combines real and virtual, infosphere and nature, data and critical thinking, information and empathy. It employs mindfulness, nature, humanity, and transdisciplinarity to oppose digital-induced intellectual laziness. We argue that a data-driven design should concretely participate in fostering an advanced digital age, building tangible actions for a renewed balance with nature.

Crafting Ecosystems is a design system that uses data to model and visualize life's intricate forms and logic. Specifically, this system focuses on two metaphorical organisms—an alga and an echinoid—and offers an alternative perspective on their unique behaviors.

The project is constituted by 3D elaborations, web-generated influences on natural morphologies, algorithm design, study maps, and silkscreens, which are purchasable in exchange for offsetting CO2 in the environment.

Specifically, the Algae experiment develops an algorithm that, drawing on the APIs of several European news outlets, analyzes media streams characterized by keyword frequency. The data determine the growth of an algae that organically transforms these streams through their repetition and interconnection.

Instead, the Echinoid experiment visualizes a real-time video where an algorithm processes tweets issued by five international governmental accounts, where the dots orient and shape the length and morphology of a sea urchin's spines. Inspired by urchins' phototaxis in response to external stimuli, the organism is deformed and grows infinitely, shaped by data and digital human communication flows.

As we have explained before, by interweaving philosophical theories with transdisciplinary design practices, the topic of echinoids and algae is examined from a perspective of interspecies coexistence. Binding agents and "entanglement" between human and nonhuman, biological and synthetic, physical and digital, the project erodes the anthropocentric view, paving the way toward a new ontological paradigm. Plant and animal species and human digital streams dialogue, transiting from one realm to another: a conversation that takes place on an earthly plane, in a kind of entanglement between entities that are only seemingly distant and noncommunicating. The planet emerges as a subject in its multiplicity of consistencies: humans, animals, plants, microorganisms, and products of technology are all "agencies," [11] entities in a hybrid network in which each contributes equally to shaping reality.

The project is based on a dual temporal perspective made evident during the pandemic lockdowns: the speed of media information from continuous digital streams and the slowness of natural ecosystems with their calm and inexorable evolutionary times. We combine these two temporal levels to relate humans and nature through digital technologies. Parallel to the speed and frenzy of the human world, there are, in fact, other worlds, other realities—slower realities with different rhythms and systems of survival. "The perception of their silent existence comes when we slow down, take the time to breathe, listen, observe, and understand what is hidden behind and within the folds of the earth" [12].

It is about two put worlds, the human world, and the vegetal or animal realm, confronted through a nonhuman dialogue that lays bare implicit relationships. Twinning is not a relationship of physical or genetic similarity but rather a relationship that two more beings who have shared the same cosmic twin mother have. "All beings that belong to the same species are twins, but all species are also twins: humans, ants, oaks, cyanobacteria, viruses are nothing more than heterozygous twins that incessantly duplicate the reality of the world whose body and mind they are" [13].

As the theories of the posthuman [14], post-digital [15], and post-anthropocentrism [16] progress and intertwine, design culture opens up new scenarios that influence the practices of numerous artists and designers—reversing the logic of massified production, human and nonhuman agencies—such as digitality, atmosphere, light, and other living organisms—can spill over into any dress or artifact, creating transformations and interactions, expressions of a multiverse existence.

Post-anthropocentrism, therefore, situates design in a pluriverse dimension: The biosphere emerges in its diversity of textures and passing states, while the matter is expressed through complex cosmic assemblages, human and nonhuman together, where differences dissolve into a plural and indistinct magma. The materials of design are closely dependent on the transdisciplinarity between heterogeneous knowledge, skills, and theoretical frameworks, and they change over time, moving from the solid state to the liquid and gaseous state to hybridization with digital flows. The dematerialization of design through digital also contributes to a new materiality. Hybrid and post-digital, the resulting aesthetic of this project is synesthetic, vibrant, and multisensorial.

5.3 Handing Data in the Digital Shift

We stand at a pivotal juncture, facing the imperative to give up blind faith in technology and cultivate our critical faculties. The global network's mission to foster equality among its users faces formidable challenges on two primary fronts: unequal accessibility to the network and its content and the ability to decipher and make informed judgments regarding the integrity and significance of disseminated information. The landscape of internet usage, intertwined with global poverty and the absence of democratic governance, manifests asymmetries in information distribution, leading to global disparities. Simultaneously, increasing reliance on machines to process information has led to a noticeable erosion of our critical thinking and discernment abilities [17].

Recognizing the significance of this juncture, the United Nations has convened a consortium of international experts to propose strategies for improving sustainable development data. Emphasizing that data forms the foundation of decision-making and accountability, the UN underscores how the current state of data underlines the gaps between knowledge acquisition and dissemination and between who is informed and who is not [18]. The vast and unconscious amount of daily digitally generated data is extensively harnessed for profit maximization and latently becomes the structure for a subtle shaping of our intrinsic desires. This amalgamation of economic and political influence exerts formidable sway, with consequences resonating through the layers of social justice. This apparent democratization of knowledge access, once in the "dreams of the digital insurrection" [19], has unintentionally fostered an ecosystem where disparities between the affluent and marginalized are transforming into a schism between the included and excluded. This asymmetry casts a shadow over entire demographic segments, even in developed nations, particularly those lacking the tools and expertise to navigate the digital landscape effectively. This is the transversal theme of media literacy; it transcends poverty, subtly eroding cognitive capacity to engage with contemporary issues and navigate the complexities of our new reality. This said, efficiently comprehending the nuances of data as an emerging material and its associated attributes assumes paramount importance. Pertinent topics, magnified during the pandemic, encompass concerns regarding information reliability and the people's capacity to discern all the presented information critically.

A fully digitalized environment, like the one we live in, has been marked by the introduction of a new communication layer into our social framework [20]. This connection gains resonance as we delve into the endeavor to reassess human existence, intricately intertwined within the tapestry of informational organisms, interlinked with biological entities and engineered artifacts [21]. We find ourselves embedded in an epoch where communication technologies wield a formidable influence, reshaping our surroundings and giving rise to a novel reality. Yet, the extent to which the field of design comprehends the intricate rules governing the ceaseless flow of data shaping our daily lives remains a pressing question. Addressing this inquiry becomes increasingly relevant, considering the profound impact exerted by these data-driven entities on the ecosystem. In our contemporaneity, knowledge has evolved into a form of currency, with information flows and data serving as its primary conduits. Information becomes produced and fundamentally commodified, serving as a wellspring for generating economic value. The outbreak of the COVID-19 pandemic laid bare vulnerabilities within our information systems, underscoring the imperative for the design discipline to engage with this transformative tide actively. The resulting necessity lies in conceiving and constructing products that seamlessly coexist within the digital era. Throughout the past century, communication has pushed human evolution and its phenomena at an unprecedented pace. Product design has continually adapted to this acceleration, seamlessly aligning with evolving needs and behavioral paradigms.

Data have become the "most important single material" [22] used in manufacturing. With its expanding spectrum of forms, velocities, and qualities, data has made significant strides, disseminating its influence widely in every aspect of life and instigating profound transformations in behaviors and beliefs. In this way, data technologies have assumed a central role in enhancing productivity, unlocking potential, and forging pathways to value creation. In such a narrative, designers have the profound responsibility to contribute to the interpretation of changes and to evaluate and propose tools for its comprehension.

The experimental project *Crafting Ecosystems* seeks to redefine the conventional role of designers about data, fostering a new understanding of dynamic and innovative design scenarios. The project has evolved through two different stages and results. The first one, *Crafting Ecosystem: Algae,* is connected with the information data and flows during the pandemic, driven by the compelling need to navigate the overwhelming infosphere of digital newspapers differently. The second one, *Crafting Ecosystem: Echinoid,* is a more mature approach that explores a different source of information and trust, more coherent with contemporaneity and its communication logic.

5.4 Crafting Ecosystems: Algae

The initial step for this project was to select a set of data sources. Starting during COVID-19, we focused on pandemic-related information through primary Italian journalistic daily news data from an open-access API. Collaborating with a network engineer, we developed an incremental set of simple algorithms intending to craft these data flows. The first algorithm filtered and retained data relevant to the prevailing COVID-19 scenario by cleaning it from redundant grammar elements nonrelevant to the scope of the research. Subsequently, we compiled a collection of frequently used words in national newspapers, spanning health, science, technology, and politics sections. We analyzed their frequency and precise usage to enable comparison, statistics, and prioritization. This part has been crucial because terminology and topics are usually treated transversally among different information areas. This data compilation provided insights into the dominant information landscape during the peak of the virus spread in Italy. One of those is that nowadays, all the classical information sources are less relevant than before, and many different communication channels cover a source of truth. Impressive, for instance, is the trust and the relevance of information delivered throughout gossip channels. Following this phase, a second algorithm was devised to extract sub-data clusters, including recurring words associated with the primary ones. This effort aimed to construct a comprehensive scenario comprising quantitative and qualitative information, creating a narrative enriched with data and perceptions. These efforts culminated in a daily collection of curated data, which was then processed using a 3D modeling algorithm to create an active data visualization medium. This transformative step envisioned the creation of organic and virtual algae, generated through the interaction of selected data and strategically filtered to convey the connection between technology, design, and ecosystem perception. To illustrate this relationship, we conceived an interactive app in which the plant's monthly growth virtually represents the narrative directions of the information system.

The result is a project in which the data algorithm governs the virtual plant's growth logic. The critical thinking project coincides with decision-making phases that determine operations at various levels of complexity (numeric, geographic, climatic data, XYZ variants). Furthermore, the plant's variations adhere to rules that are only partially controllable due to their ties to media information flows and atmospheric parameters, emphasizing the dynamic nature of this digital ecosystem. The visualization via the app on digital devices seeks to transform readers into active participants in the daily ecosystem, fostering empathy and interaction with the variability of the information system (Figs. 5.4 and 5.5).

5.5 Crafting Ecosystems: Echinoid

Starting with a more mature and conscious understanding, the research intention was to explore the information flows through a more contemporary medium. The information system is mainly characterized by speed, synthesis, predominance of loud headlines, strong statements, compulsive sharing, and active deepening. The selected data source, in this case, has quickly been found on Twitter (now X). The platform contains all the above characteristics and, at the moment of the research, had a reliable resource page and accessible open API for our research purposes. The intention to continue working within the metaphor of nature has brought us to evaluate the most fascinating natural behavior to address. In this sense, echinoids represent a fascinating biological entity. They belong to the phylum Echinodermata and

Fig. 5.4 The illustration represents the final algae modeled by the selected API open access data

the class Echinoidea. They have a hard shell covered with small knobs to which spines are attached in living echinoids. Their main characteristic is that they develop into a five-fold symmetry. Mostly, echinoids have no brain or heart. Despite this, they demonstrate a very complex form of interacting with the environment they live in. Research has proved that their movements are all but scopeless. Phototaxis is the central paradigm with which they explore their marine environment, meaning that the light sources in their surroundings inform their directions. This movement, which is not controlled by a central system, is influenced mainly by external

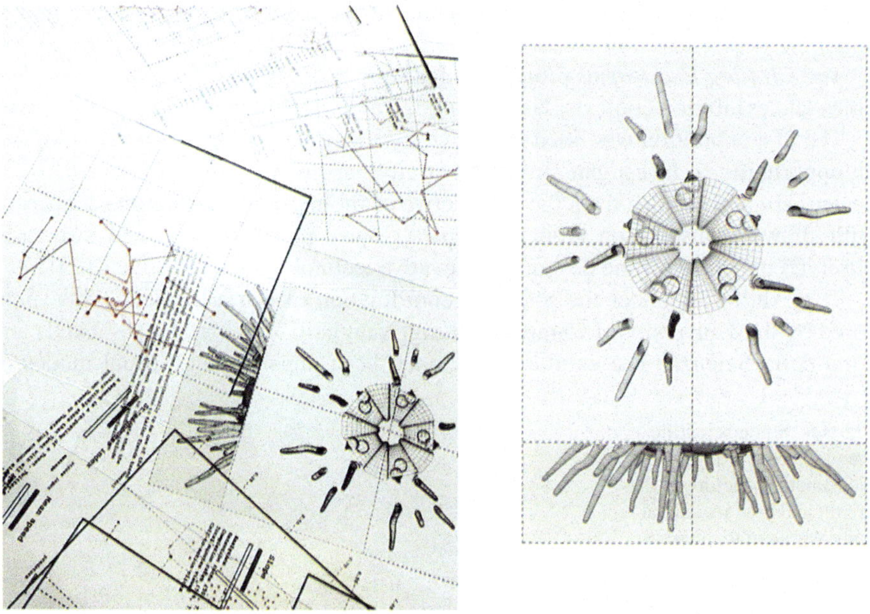

Fig. 5.5 This illustration represents the functionality of the app and the different screens appearing until the purchase moment

elements, such as other organisms' mechanical disturbance, chemical stimuli, predator perception, and tides.

The biological reference is highly accurate compared to how a communication medium like Twitter functions. It has, in fact, a multi-folded hyper-symmetric structure not controlled by any central brain, which is influenced in its vitality by external interactions of the digital environment. More precisely, Twitter information flows are influenced and spread under the contemporary rules of likes, retweet, and answers. Within these interactions, the information system evolves in unpredictable directions and consequences. For the sake of the research, we have then decided to focus on a restricted group of Twitter accounts with political and global influential criteria. We have then created an algorithm to track the flows of information delivered through five official political accounts of global state leaders. Five as the sea-urchin symmetrical structure. Using the same behavioral scheme of the animal, the algorithm is constantly detecting, for each account and each tweet, the environmental influences that constitute the structure for their development and spread in the platform. Likes, answers, and reposts dynamically inform the construction of this flow. As a metaphor for representation, we have decided to look again at the sea urchin, creating an artificial animal whose spines represent the development of a single tweet evolution from each account in time. Due to the tweet's scores, every spine evolves in length, girth, and direction. This creates an asymmetrical, nonnatural, deformed virtual animal that metaphorically represents the impact of unstructured information on our lives and environment. The amorphous virtual sea-urchin becomes the fetish of an unbalanced and uncontrolled information system (Figs. 5.6 and 5.7).

The *Crafting Ecosystems* projects—algae and echinoids—were exhibited in two cities and exhibition contexts: Naples in 2021 and Florence in 2022.

The first exhibition was titled *EchinoDesign*, an international exhibition that was an opportunity to investigate, in a transdisciplinary way, the relationship between design, art, and science from the perspective of biomimicry, a discipline concerned with drawing inspiration from principles, logic, morphologies, and structures observed in nature for the design of innovative technologies and artifacts [23].

The exhibition, under the scientific coordination of Carla Langella, was sponsored by the University of Campania "Luigi Vanvitelli" (Department—DADI) and Città della Scienza. The exhibition included both physical and virtual modes of

Fig. 5.6 Some scientific drawings concerning the biomimetic structure of the urchin and how its animal body works in the marine space

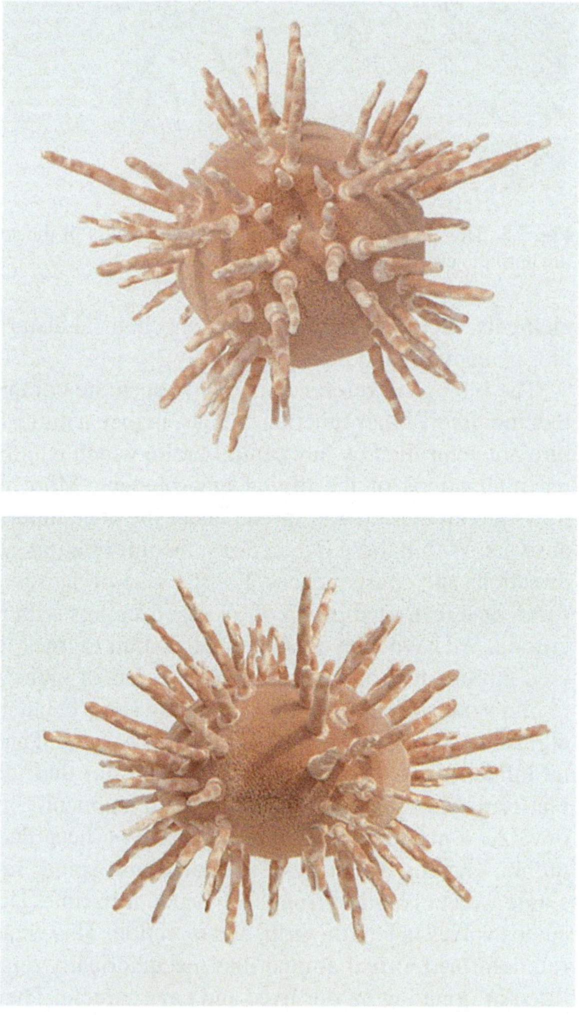

Fig. 5.7 Sea urchin model by two different perspectives

fruition. It included design, artistic, and scientific products that interpreted, biomimetically, the biological characters of echinoids (sea urchins) observed from different points of view and through varying levels of investigation—morphological, generative, structural, material, dynamic, and compositional. In this context, bordering on the natural sciences, *Crafting Ecosystems* was exhibited through an installation table where different types of media, including real urchins, photographs of ancient drawings, silkscreens, and scientific data, were interwoven to create a sophisticated transmedia narrative (Fig. 5.8).

From November 3 to 24, 2022, the project was exhibited at the fifth edition of God Is Green, the festival dedicated to sustainability, conceived by Manifattura Tabacchi, with the collaboration of DIDA—Department of Architecture of the University of Florence and NAM—Not A Museum [24]. The exhibition, titled *Reverse Towards a New Paradigm*, displayed installations, videos, and documentation works, developing a narrative path to raise awareness of the current state to reconstitute a new balance between human and natural.

Focusing on climate, social, and political emergencies, the curatorship, coordinated by Prof. Giuseppe Lotti, involved 17 designers characterized by a particular speculative and unconventional slant. The exhibition itinerary was developed in five chapters: starting with a rupture of natural boundaries, traversing the concepts of mutual altruism and technology-oriented to innovative solutions, and ending with the idea of tracing in the cosmos a possible way to reconnect. As part of *Reverse Towards a New Paradigm*, the *Crafting Ecosystems* project tackled the topic of technology with purpose, proposing technological innovation as a narrative tool to tell

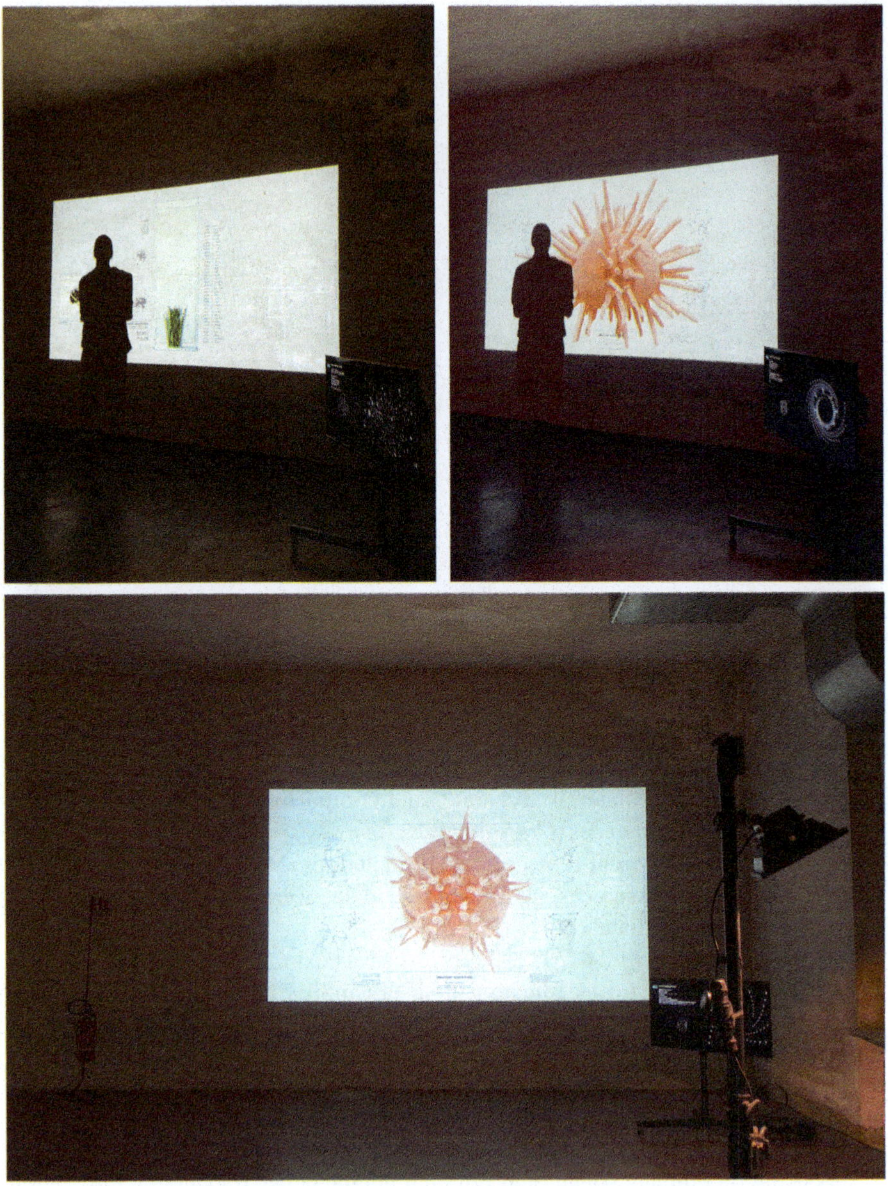

Fig. 5.8 The photographs represent the *EchinoDesign* exhibition at Città della Scienza in Naples. In this exhibit, all the images, graphs, and ancient selected pictures were exhibited on one unique table, together with the ancient natural urchin that has bioinspired the 3D further video modeling

Fig. 5.9 *Crafting Ecosystems installation at the Reverse Towards a New Paradigm* exhibit in Florence: two 3D videos, one algorithm video in real time, and two mixed media frames

the environmental impact of our actions. For this exhibition, the project created two video installations that, in real-time, demonstrated the dialogical relationship between the two living organisms—seaweed and sea urchin—and an algorithm that modeled them through different tweets drawn from the media streams of mass information (Fig. 5.9).

Because of its ability to illustrate complex issues through a data visualization activity, *Crafting Ecosystems* is a particularly relevant and incisive experiment in its storytelling of emergent instances aimed at an ecosystemic awareness toward other species and realms. The above design process seeks to create active short circuits to change how designers can manipulate matter of data to build their artifacts.

How can new forms of technology and design culture promote ecological awareness and bring about real change in our consumption patterns? To envision a future that safeguards biodiversity for all living beings, we must establish relationships beyond the human realm and embrace diversity.

This paper offers an implementation of a hybrid and dialogic system to foster a connection between living beings and data. The aim is to encourage discussions and to develop truly impactful ecological behaviors by combining human and nonhuman data in new and innovative ways (Figs. 5.10 and 5.11).

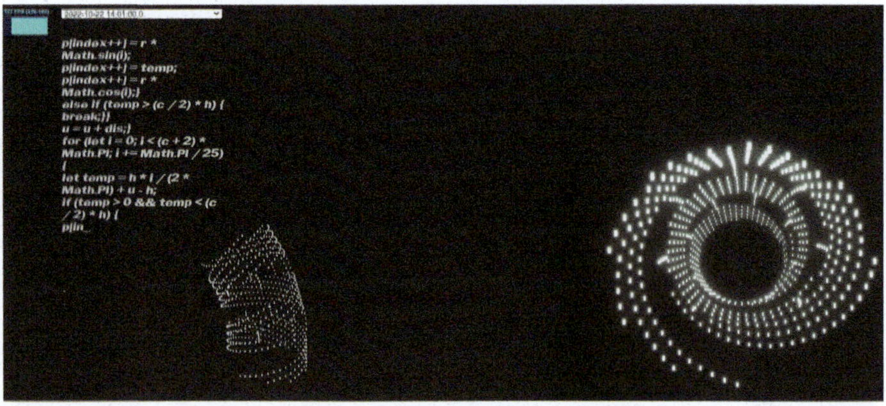

Fig. 5.10 These photographs represent the algorithm's movement through some frames. All the spine's points came from the single input of a tweet account. The official Biden's Twitter account has been used for this specific frame

Fig. 5.11 These photographs represent the algorithm's movement through some frames. All the spine's points came from the single input of a tweet account. The official Biden's Twitter account has been used for this specific frame

5.6 Conclusions

Crafting Ecosystems is an experiment that uses data visualization and design simulation to explain complex issues of nature-culture-society entanglement. It contributes to the definition of alternative storytelling of emergent problems by design, aiming to create awareness of other entities and realms. Through a speculative approach, this design research wants to trigger short circuits to change how designers can manipulate data and create artifacts [25]. One of our questions is how design culture can use new technologies and biomimicry to promote ecological awareness and sustainable consumption [26].

To create a future that prioritizes biodiversity protection for the benefit of all living beings, we need to build relationships that go beyond human interactions and embrace diversity. The coexistence of humans and nonhuman entities can be realigned by mixing them in the design process and products to stimulate a debate on truly impactful ecological behaviors.

In this proposal, raw data is seen as a physical material that can convey information and model a tangible change process. The goal is to use digital technologies to connect humans and nature and observe reality in an integrated manner. By doing so, we can understand how emergencies depend on the interconnected balance between all living beings, including humans, animals, and plants. *Crafting Ecosystems* hybridizes data, algorithms, and social media dynamics with biological information, behaviors, and visual representations. Virtual entities exist, function, and grow in the digital realm with features, proportions, and performances of natural creatures to challenge human beings' perception, identity, and belonging.

Acknowledgments We thank ZHANG Rengzheng, a student from SUSTech School of Design, for his work on the algorithm for the project Crafting Ecosystems: Echinoids and Paolo Di Grazia and for his support for the 3D modeling organisms.

References

1. Zizek S (2010) Welcome to interesting times! (2012). Benvenuti in tempi interessanti. Ponte delle Grazie, Milan
2. Escobar A (2018) Designs for the pluriverse. Radical interdependence, autonomy, and the making of worlds. Duke University Press, Durham
3. Morton T (2013) Hyperobjects: philosophy and ecology after the end of the world. University of Minnesota Press, Minneapolis
4. Morton T (2021) Ecologia oscura. Logica della coesistenza futura. Luiss University Press, Rome
5. Benasayag M (2022) Corpi viventi: pensare e agire contro la catastrofe. Feltrinelli, Milan
6. Haraway DJ (2016) Staying with the trouble: making kin in the Chthulucene. Duke University Press, Durham
7. Scarpitti C, Dell'Aglio F Migliore E (2021) Crafting ecosystems with data. The value of design in the emergency-driven scenarios. In: Di Lucchio L, Imbesi L, Giambattista A, Malakuczi V (eds) Design culture(s). Cumulus conference proceedings Roma 2021, vol. #2. Cumulus Publisher, Aalto, pp 2537–2550
8. Piscitelli D (2019) First thing first. Comunicare le emergenze. Listlab, Barcelona
9. Antonelli P, Tannir A (2018) Broken nature: design takes on human survival. Electa, Milan
10. Lowenhaupt Tsing A (2015) The mushroom at the end of the world: on the possibility of life in capitalist ruins. Princeton University Press, Princeton
11. Latour B (2005) Reassembling the social: an introduction to actor-network-theory. Oxford University Press, Oxford
12. Caponetti F (2019) Terre Marginali. Agricoltura come nuovo Umanesimo. Quodlibet Studio, Macerata

13. Coccia E (2018) La vita delle piante: Metafisica della mescolanza. Il Mulino, Bologna
14. Braidotti R (2014) Il Postumano. La vita oltre l'individuo, oltre la specie, oltre la morte. DeriveApprodi, Milan
15. Alexenberg M (2011) The future of art in a postdigital age. Intellect, Bristol
16. Escobar P et al (2021) Pluriverso. Dizionario del post-sviluppo. Orthotes
17. Perez Tornero JM, Varis T (2010) Media literacy and new humanism. UNESCO Institute for Information Technologies in Education, Moscow
18. United Nations (2014) A world that counts mobilising the data revolution for sustainable development (Report). http://www.sustainabledevelopment.un.org
19. Baricco A (2018) The game. Einaudi, Torino, p 169
20. Toffler A (1980) The third wave. William Morrow, New York
21. Floridi L (2015) The onlife manifesto: being human in a hyperconnected era. Springer, Oxford
22. Duncan LW (1994) Manufacturing 2000. Amacom, New York
23. Echino Design. https://www.hybriddesignlab.org/exhibitions/echinodesign/
24. Reverse Towards a New Paradigm. https://www.manifatturatabacchi.com/en/eventi/reverse-towards-a-new-paradigm-god-is-green-v-edition/
25. Dunne D, Raby F (2013) Speculation everything. MIT Press, Cambridge
26. Zeng L (2023) Nathaniel Tkacz, being with data: the dashboarding of everyday life. Int J Commun 17. http://ijoc.org

Francesco Dell'Aglio is an industrial designer and holds a PhD with a background in Architecture. He teaches Project Development and UI/UX Design at Apple Developer Academy in Naples. His research interests are culturally connecting materials and manufacturing processes and investigating relationships with the environment and its perception.

Enza Migliore is an assistant professor at SUSTech School of Design as the PI of the Materialities Research Group. Her work explores materials through interdisciplinary collaboration between design, materials science, art, and experimental and speculative design practices. She tackles engagements in material culture and commitment to designing with ecological perspectives.

Chiara Scarpitti is a designer and holds a PhD in Design and Innovation. She is an assistant professor at the Department of Architecture and Industrial Design, University of Campania "Vanvitelli," Italy. Her research activities—theoretical and practice-based—are characterized by a speculative hybridization between digital technologies and high craftsmanship in a vision of post-digital convergence.

Chapter 6
Design Beyond the Environmental Pillar: Potential Contributions of Biomimicry to Social Design

Luiz Valdo Alves Maciel Filho and Amilton José Vieira de Arruda

Abstract Biomimicry is a fascinating approach that seeks inspiration from nature to solve human problems and create sustainable solutions. It is based on the idea that nature, over billions of years of evolution, has developed effective solutions for a wide range of challenges, and we can learn from these efficient designs. Meanwhile, Maslow's Hierarchy of Needs is a psychological theory that describes human needs in a hierarchy, from basic needs to self-actualization needs. In this chapter, we will explore how biomimicry and Maslow's theory can intersect, providing an interdisciplinary approach to sustainable innovation.

Keywords Biomimicry · Sustainability · Social design · Maslow's Pyramid

6.1 Introduction

Many scholars have highlighted biomimicry as a way to seek sustainable solutions to human challenges, including optimization of forms and processes, material efficiency, and energy consumption. However, it is important to note that many studies involving sustainability tend to overly focus on environmental and economic aspects, sometimes neglecting the social aspect. As Gbededo [1] has already pointed out in their research, many studies aimed at understanding and improving sustainability predominantly concentrate on these areas, with only 8% of the entire body of research showing concern for the social issues and impacts that communities experience during the development of new technologies, products, and services.

This study aims to fill this gap by seeking to understand biomimicry and its analogs from the core of the design methodology. Believing that it inherently embodies

L. V. A. Maciel Filho (✉) · A. J. V. de Arruda
BioDesign Lab, Programa de Pós-Graduação em Design (PPGDesign), Federal University of Pernambuco (UFPE), Recife, PE, Brazil
e-mail: luiz.valdo@ufpe.br; amilton.arruda@ufpe.br

A. J. V. de Arruda, F. L. Palombini (eds.), *Biology, Biomimetics and Natural Design*, Environmental Footprints and Eco-design of Products and Processes,
https://doi.org/10.1007/978-3-031-65877-8_6

ethical principles aimed at social well-being and improving the living conditions of the population. To guide this quest, we will use Maslow's Hierarchy of Needs [2] as a model to measure social well-being, identifying at which level of the pyramid, individual and community needs have the potential to be met.

The methodological procedures will involve qualitative analysis conducted through a literature review of biomimicry, comparing its objectives with the different levels of Maslow's Pyramid [2], and relating the two to assess the fulfillment of these needs this methodology has been previously applied by Maciel Filho et al. [3] in their study concerning the social impact that endeavors focused on the sustainability tripod could achieve. The analysis consists of a literature review of the concepts and understanding of Maslow's theory [2], which is subsequently compared with the results obtained by the endeavor.

In an adaptation for application to biomimicry, two concepts and theories will be compared to prepare for future studies that can be applied in cases where biomimicry can be implemented.

6.2 Striving for Self-Actualization: A Revisited Maslow's Theory

Maslow's Hierarchy of Needs Theory, developed by Abraham Maslow in 1943, continues to be a cornerstone in understanding human motivation and personal development [2]. Initially comprised of five tiers, the theory has evolved over the years to include eight tiers that represent human needs in a pyramid. This pyramid represents each individual's journey of self-discovery and self-actualization throughout life.

Previously, it was believed that people sought to fulfill the needs at the base of the pyramid before aspiring to the higher layers. These basic needs include physiological, safety, belongingness, esteem, and self-actualization needs [2]. However, the theory was updated to reflect a more flexible view of human development, allowing needs to be pursued simultaneously.

The eight tiers, classified as D-needs (deficiency needs) and G-needs (growth needs), encompass essential areas of human existence. They include physiological, safety, belongingness, esteem, cognitive, aesthetic, self-actualization, and transcendence needs [4]. Each of these tiers plays a fundamental role in the pursuit of self-actualization and personal growth.

Physiological needs: At the base of the pyramid are physiological needs, which encompass essential requirements for human survival, such as food, clean water, shelter, and proper rest. When these needs are unmet, there can be severe harm to both physical and mental health, making them the primary motivators [5].

Safety needs: Just above physiological needs are safety needs. This involves seeking protection against imminent threats, whether in the form of physical, legal, or financial security. Lack of safety can lead to anxiety and hinder an individual's ability to plan for the future with clarity [5].

Belongingness needs: At the third level of the pyramid, we find the need for belongingness. This refers to the inherent desire of human beings to connect with others, whether for teamwork, emotional expression, or idea sharing. The absence of interpersonal connections can lead to psychological issues, highlighting the importance of social relationships in human life [6].

Esteem needs: The need for esteem, one of the deficiency needs, is subdivided into self-esteem and esteem received from others. Self-esteem is crucial for progressing to the higher tiers of the pyramid, as it fosters confidence and self-sufficiency. While the esteem of others is relevant, self-esteem plays an even more crucial role in self-actualization [4].

Cognitive needs: This tier involves the relentless pursuit of knowledge, intellectual curiosity, exploration of new ideas, and the need for predictability. These are essential for intellectual development and the enhancement of cognitive skills, intrinsically linked to the quest for meaning and understanding of the world around us.

Aesthetic needs: Aesthetic needs pertain to the appreciation of beauty, balance, and harmony in various aspects of life. They involve the search for the sublime, both in art and nature, reflecting the human need for aesthetic experiences that inspire and enrich the soul.

Self-actualization needs: At the top of the pyramid lies the need for self-actualization, where individuals seek to reach their full potential and become the best version of themselves. Characteristics associated with self-actualization include creativity, autonomy, resistance to indoctrination, spontaneity, and the continuous pursuit of personal growth [4]. This category applies to all individuals, not just notable people.

Transcendence needs: The pursuit of transcendence involves values that go beyond personal experience, encompassing mystical, religious, sexual, scientific, and aesthetic aspects. It is the need to connect with something greater than the individual self, experiencing ecstasy in spiritual, contemplative, or aesthetic experiences that transcend the boundaries of common existence [4].

The most up-to-date configuration of Maslow's Hierarchy of Needs pyramid can be represented as shown in Fig. 6.1.

The influence of social and cultural context is undeniable in meeting these needs. Meeting physiological needs takes priority, as it is directly linked to survival. When this base is not satisfied, other needs become secondary.

In summary, Maslow's Theory offers a comprehensive framework for understanding human motivations and personal development [2]. Its eight tiers highlight the complexity of the journey towards self-actualization and how human needs can be flexibly satisfied. Social and cultural context plays a significant role in this journey of self-discovery and personal growth.

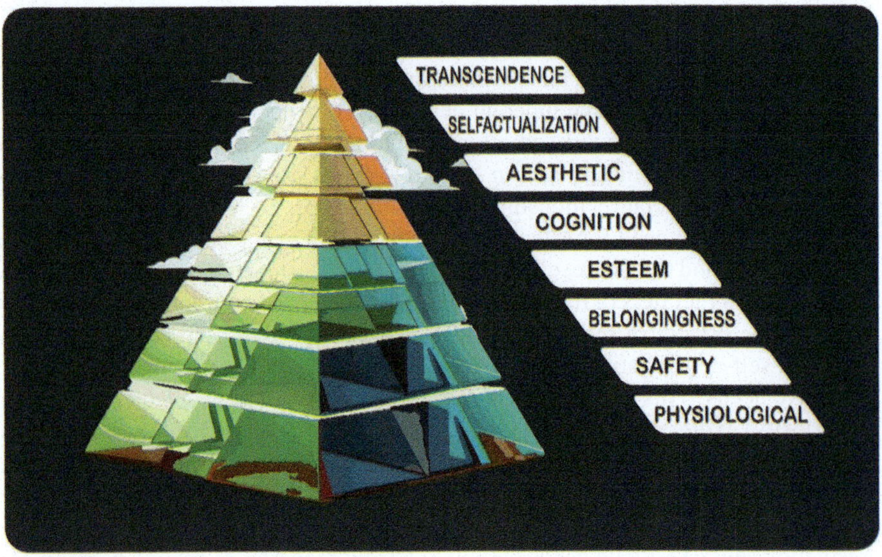

Fig. 6.1 Maslow's hierarchy of needs pyramid. (Image interpreted by the author, CC-BY License)

6.3 Biomimicry and Its Importance for Sustainability: A Social Approach

Nature, once revered as divine, has lost its symbolic status, becoming a mere source of natural resources [7]. Along with this symbolic loss of "nature," we have also lost the quality of life and contemplative time, despite technological revolutions that promised more free time for individuals. Instead, we have witnessed an excessive proliferation in the production of consumer goods and the creation of false needs [8].

In this context, biomimicry emerges as a field of knowledge that offers solutions to human challenges, especially in the creative sphere, based on biological studies [9, 10]. Biomimicry recognizes nature as a model, a measure, and a mentor, incorporating its inspirations, effective solutions, and the ability to inspire new perspectives on the biosphere [9, 10].

Within biomimicry, several terms are crucial:

Bioinspiration: This term is widely used in biomimetic design, referring to inspiration drawn from organisms, both in their morphology and behavior, and how they interact with their ecosystems [11].

Biomorphism: This concept involves the visual and stylistic reproduction of natural forms, as often seen in the Art Nouveau movement.

Biophilia: Referring to the emotional connection between humans and nature, biophilia can promote performance improvements and health benefits [12].

Bioutilization: Involves the use of natural resources in their original form, such as using tree leaves as roof coverings or wood for construction.

Bioassistance: Occurs when organisms are domesticated to fulfill a specific function, such as cultivating plants to repel insects or using vegetative coverings on facades to improve the thermal sensation of environments [11].

Biotechnology: This term describes the processes that humans develop based on the observation of natural structures, including concepts like bionics and biodesign [13].

Bionics: In the field of design, bionics seeks to prioritize the application of the formal, functional, material, and evolutionary characteristics of nature in technical, mechanical, and industrial systems, without the need for direct integration with the natural environment [13].

Biomimicry: This concept represents a synthesis between human inventiveness and effective solutions found in nature. It goes beyond the simple integration of natural elements into designs, seeking analogies and abstractions to understand how nature achieves its results and to reproduce them [11].

Biomimicry, therefore, is characterized as a field of knowledge that studies natural forms, processes, and systemic relationships for human production, resulting in positive impacts on sustainability. It proposes solutions adapted to the local context, uses non-toxic elements, and promotes the efficient use of resources [11].

As Benyus [9] highlights, after billions of years of evolution, the natural system has developed structures and functions adapted to each context, especially when it comes to sustainable and efficient systems. Biomimicry recognizes that the natural environment is not just a source of resources to be preserved for future generations but also a repository of effective and sustainable best practices that can guide humans in creating solutions for human, social, and collective well-being.

To ensure the ethical application of biomimicry, internal changes in humans are necessary, including adopting responsibility for the biosphere and sensitivity to learning from nature [14].

However, it is encouraging to see that awareness of the importance of preserving nature is growing in many populations, as demonstrated by the updated study by CJF, comparing German perceptions of nature preservation with those of the rest of the world.

Papanek [15] already encouraged observing nature to guide projects, going beyond the inclusion of visual elements and investigating biological principles and the interrelationships of its parts and systems. This multidisciplinary approach can enhance the reconnection of humans with the biosphere [15, 16].

In summary, biomimicry is a powerful tool that can significantly contribute to sustainability, encompassing economic, environmental, and social aspects. Of course, this will only be possible when humans make an effort to understand and learn from nature.

6.4 The Intersection Between Maslow's Hierarchy and Biomimicry: Implications for Personal Ascension and Social Achievements

Maslow's Hierarchy of Needs, a widely recognized psychological theory, outlines human needs in a hierarchy that spans from the most basic to the highest. In parallel, biomimicry, a field of study that seeks solutions to human challenges by observing and imitating nature, stands out as a promising approach to meet these needs. Examining the intersection between Maslow's Hierarchy and biomimicry reveals how this combination can drive personal ascension for individuals and contribute to significant social achievements.

Physiological needs and biomimicry: Physiological needs, which form the base of Maslow's Hierarchy, include essential elements such as food, water, and shelter. Biomimicry plays a crucial role here, as nature offers numerous effective solutions for survival. By observing how organisms in nature obtain food and water [17], as well as build resilient shelters, biomimicry can inspire designs that ensure sustainable access to basic resources [18]. Projects incorporating strategies like rainwater harvesting, water biofiltration, and efficient food production reflect this approach, promoting the security of physiological needs.

Security needs and biomimicry: Security needs in Maslow's Hierarchy involve not only physical protection but also emotional and financial security. Biomimicry can contribute to this area by inspiring solutions that mimic natural systems of protection and adaptation. For example, studying how animals build their shelters to withstand natural disasters can inform resilient construction projects. Furthermore, nature provides examples of adaptive systems that can be applied in cybersecurity and finance, aiding in threat detection and response. By adopting these biomimetic approaches, we can strengthen security in all its dimensions.

Affiliation needs and biomimicry: The need for affiliation refers to our search for interpersonal relationships and belonging to social groups. Biomimicry, by focusing on interactions between organisms and ecosystems, can offer valuable insights into cultivating healthy and constructive relationships. Models based on symbiosis and collaboration in nature can be applied to promote greater understanding, empathy, and harmony in human relationships [19]. Additionally, by observing how animal groups function cooperatively, we can learn to improve collaboration in teams and communities.

Esteem needs and biomimicry: Esteem needs include both self-esteem and recognition and respect from others. In this context, biomimicry can inspire the building of self-esteem by highlighting how organisms in nature adapt, thrive, and contribute to their ecosystems. Observing how nature values diversity and individual contributions can nurture a sense of self-worth in humans. Furthermore, biomimicry can also influence how we perceive and respect others, emphasizing the interdependence and interconnectedness of all living beings.

Cognitive needs and biomimicry: Cognitive needs represent the search for learning, knowledge, and curiosity. Biomimicry, with its focus on bioinspiration, biomorphism, and bioutilization, stimulates human curiosity and the pursuit of knowledge. It also guides the design of effective educational environments [16], leveraging efficiency and adaptability principles found in nature. Through biomimicry, education can become more engaging and aligned with the cognitive needs of individuals.

Aesthetic needs and biomimicry: Aesthetic needs include the appreciation of beauty and balance. Biomimicry influences aesthetics by inspiring an appreciation for natural and artistic beauty (bioinspiration). Additionally, it guides the design of aesthetically pleasing spaces (biomorphism and biophilia), enhancing the aesthetic environment for well-being. The pursuit of beauty in nature can inspire the creation of environments that promote a deeper aesthetic appreciation, contributing to the fulfillment of individuals' aesthetic needs.

Self-actualization needs and biomimicry: At the top of Maslow's Hierarchy is the need for self-actualization, which involves the pursuit of personal growth and the realization of human potential. Biomimicry, with its emphasis on nature-inspired innovation, offers a path to fulfill this need. By studying examples of creativity, autonomy, resilience, and acceptance found in nature, we can be inspired to achieve our higher personal goals. Furthermore, the pursuit of sustainable solutions through biomimicry directly contributes to self-actualization, as it is often linked to creativity and innovation.

Transcendence needs and biomimicry: The need for transcendence represents the search for meaning, purpose, and connection to something greater than the self. Biomimicry, by recognizing the interconnection and interdependence of all living beings in nature, can inspire a profound sense of connection to the natural world. Moreover, by seeking nature-inspired solutions that promote sustainability and collective well-being, biomimicry contributes to fulfilling this need for transcendence. By applying biomimetic principles to create a more sustainable and harmonious future, individuals can find a greater purpose in their lives and contribute to significant social achievements.

From the analyses conducted, it is possible to create a synthesis framework that facilitates understanding and organization of the points that biomimicry and its analogs can address (Table 6.1).

6.5 Contributions of Biomimicry to Personal Growth and Social Achievements

The intersection between Maslow's Hierarchy and biomimicry provides an opportunity to enhance the quality of life and promote social achievements. By addressing physiological and safety needs sustainably, we establish solid foundations for personal growth. Inspired by nature, we can improve our interpersonal relationships, develop healthy self-esteem, seek knowledge, appreciate aesthetics, and realize our full potential. Furthermore, by pursuing self-actualization and transcendence

Table 6.1 How Biomimicry and its analogues can contribute to achieving Maslow's Hierarchy

Maslow's needs	Terms of biomimicry	Influence on meeting the need
Physiological needs	Bioutilization	Ensures sustainable access to food and water
	Biomorphism	Inspires designs for resilient shelters
	Bio assistance	Improves thermal comfort in environments
Safety needs	Biotechnics	Reinforces cybersecurity and financial security
	Emulate	Enables effective adaptation to threats
	Systemic emphasis	Strengthens resilience and adaptability to challenges
Need for affiliation	Biophilia	Cultivates healthy interpersonal relationships
	Bioinspiration	Promotes collaboration in teams and communities
Esteem needs	Bioutilization	Builds self-esteem through adaptation
	Biomimicry	Values diversity and individual contributions
Self-actualization needs	Emulate	Inspires creativity and the pursuit of sustainable solutions
	Systemic emphasis	Contributes to innovation and the realization of potential
	Bioinspiration	Fosters autonomy and personal resilience
Cognitive needs	Biomorphism	Stimulates curiosity and the pursuit of knowledge
	Bioutilization	Guides the design of educational environments
	Bioinspiration	Promotes learning from nature
Aesthetic needs	Biomorphism	Inspires appreciation for natural and artistic beauty
	Biophilia	Guides the design of aesthetically pleasing spaces
	Biomimicry	Enhances the aesthetic environment for well-being
Transcendence needs	Emulate	Encourages the pursuit of passions and individual achievements
	Systemic emphasis	Facilitates overcoming personal challenges
	Bioutilization	Empowers meaningful contributions to society

through biomimicry, we can contribute to more sustainable solutions in our society, fostering a profound sense of purpose and connection with the natural world and the global community.

In summary, integrating biomimicry with Maslow's Pyramid theory offers a promising path for personal and social development. Nature, with its accumulated wisdom over billions of years, can serve as a valuable guide to fulfilling human needs and achieving higher objectives. By embracing this approach, we can create a more sustainable and harmonious world where people thrive in all aspects of their lives.

6.6 Conclusions

Biomimicry and Maslow's Hierarchy, despite belonging to different fields, can be seen as two complementary lenses for understanding and addressing human needs and contemporary challenges. By integrating these two perspectives, we can create more holistic and sustainable solutions that meet human needs while respecting the ecological principles that nature has honed over millennia. The pursuit of sustainable innovation and human well-being can be driven by this interdisciplinary approach, where nature becomes our mentor and guide.

The combination of biomimicry and Maslow's Pyramid can lead to significant advances in areas such as design, architecture, medicine, and technology, helping us create a future that is more in harmony with the natural world and that satisfies human needs more comprehensively and satisfactorily.

References

1. Gbededo MA, Liyanage K, Garza R, Arturo J (2018) Towards a life cycle sustainability analysis: a systematic review of approaches to sustainable manufacturing. J Clean Prod 184:1002–1015. https://doi.org/10.1016/j.jclepro.2018.02.310
2. Maslow AH (1943) A theory of human motivation. Psychol Rev 50(4):370–396. https://doi.org/10.1037/h0054346
3. Maciel Filho LVA, Silva NoguelRA AH, Vieira De Arruda AJ, Oliveira Clementino T (2023). El uso de herramientas de fabricación digital para la innovación social y ambiental en tiempos de crisis postpandemia. Cuadernos del Centro de Estudios de Diseño y Comunicación, n. 190, 24 May
4. Mcleod S (2018) Maslow's hierarchy of needs
5. Cavalcanti TM, Gouveia VV, de Medeiros ED, Mariano TE, de Moura HM, Moizeís HBC (2019) Hierarquia das Necessidades de Maslow: Validação de um Instrumento. Psicologia: Ciência e Profissão, 39. https://doi.org/10.1590/1982-3703003183408
6. MacDonald G, Leary MR (2005) Why does social exclusion hurt? The relationship between social and physical pain. Psychol Bull 131(2):202–223. https://doi.org/10.1037/0033-2909.131.2.202
7. Escobar A (1995) Sustainable development: the death of nature and the rise of environment. In: Escobar A (ed) Encountering development: the making and unmaking of the third world. Princeton University Press, Princeton
8. de SÁ AAM (2021) Ferramentas da biomimética no design. Aportes da natureza para a prática projetual. Universidade de Brasília, Brasília
9. Benyus J (1997) Biomimicry: innovation inspired by nature. William Morrow, Estados Unidos
10. Benyus JM (2011) Biomimétisme: quand la nature inspire des innovations durables. Rue de l'Échiquier, Paris
11. Baumeister D, Tocke R, Ritter S, Dwyer J (2014) Biomimicry resource handbook: a seed bank of best practices. Biomimicry 3.8, Montana
12. Wilson EO (1984) Biophilia: the human bond with other species. Harvard University Press, Cambridge, MA/London
13. de Arruda AJV (org.). (2018) Métodos e processos em biônica e biomimética: a revolução tecnológica pela natureza. Blucher, São Paulo. Cidade MK, Perini JT, Palombini FL (2022) Bionics for inspiration: a new look at brazilian natural materials for application in sustainable

jewelry. In: Palombini FL, Muthu SS (eds) Bionics and sustainable design. Springer, Singapore, pp 195–223
14. Antonioli M (2017) Biomimétisme: science, design et architecture. Éditions Loco, Dijon
15. Papanek V (1985) Design for the real world: human ecology and social change. Thames & Hudson, London
16. Soares TLDF, de Arruda AJV (2018) Fundamentos da Biônica e da Biomimética e exemplos aplicados no laboratório de Biodesign na UFPE. In: de Arruda AJV (org.). Métodos e processos em biônica e biomimética: a revolução tecnológica pela natureza. Blucher, São Paulo, pp 9–35. Disponível em: https://www.blucher.com.br/
17. Brown PS, Bhushan B (2016) Bioinspired materials for water supply and management: water collection, water purification and separation of water from oil. Philos Transact A Math Phys Eng Sci 374(2073):20160135. https://doi.org/10.1098/rsta.2016.0135
18. Perennial Grain Crop Development (2016, agosto 30) The Land Institute. https://landinstitute.org/our-work/perennial-crops/
19. O que é a Economia Circular? ([s.d.]) Green America. Recuperado 2 de setembro de 2023, de https://www.greenamerica.org/consume-less-live-more/what-circular-economy

Luiz Valdo Alves Maciel Filho is a PhD student at the Federal University of Pernambuco (UFPE) and a researcher at the BioDesign Laboratory—UFPE. He is also a scholarship recipient in the postgraduate program by FACEPE and a volunteer in the Precious Plastic UFPE program.

Amilton José Vieira de Arruda graduated in Industrial Design Product Project from UFPE and holds a Master in Design and Bionics from IED in Milan and a PhD in Ricerca in Disegno Industriale from the University of Milan Politecnico. He also holds a postdoctoral in Design and Bionics from IADE European University UNIDCOM, Lisbon and a postdoctoral in Bionics from the University Luigi Vanvitelli Naples—IT. He is associate professor IV at the UFPE Design Course and coordinates the BioDesign Lab.

Chapter 7
Nature as Inspiration in Learning Processes

Marlén López and Manuel Persa

Abstract Taking Nature as a model makes us rethink and reimagine the future of design and manufacturing processes, through alternatives to traditional teaching methods. *SummerLab Biodesign* is an international immersive learning program in the heart of Nature, located in the Natural Park of Redes (Asturias, Spain). Through various learning modules, we offer people different methodologies, tools, and resources to learn to observe Nature with curiosity and find alternatives in the design processes through creative experimentation. *Biodesign School* is a discovery and learning program for families and children in the fields of biodesign and biofabrication. Through various activities and experimental laboratories, young ones delve into the fascinating world of creativity, science, and technology, all while being in direct contact with Nature. In this paper, we explore the transformative potential of using nature as a model to revolutionize the future of design and manufacturing processes. We delve into innovative approaches that challenge traditional teaching methods, advocating for a more immersive and experiential style of learning.

Keywords Biodesign · Biofabrication · Biomimicry · Bioinspiration · Nature-inspired design · Sustainability · Circular economy · Immersive education

7.1 Introduction

In contemporary education, there is a growing recognition of the need for innovative teaching methodologies that can effectively address the multifaceted Sustainable Development Goals (SDGs) [1]. This paper presents an exploration of how nature-inspired learning processes can serve as a catalyst for transformative educational practices. Throughout the following sections, the pressing need for innovative

M. López (✉) · M. Persa
Laboratorio Biomimético Ladines, Ladines, Asturias, Spain
e-mail: info@laboratoriobiomimetico.com; info@laboratoriobiomimetico.com

A. J. V. de Arruda, F. L. Palombini (eds.), *Biology, Biomimetics and Natural Design*, Environmental Footprints and Eco-design of Products and Processes,
https://doi.org/10.1007/978-3-031-65877-8_7

teaching methods that can effectively respond to the challenges posed by the twenty-first century is highlighted. Traditional educational approaches often struggle to equip students with the essential skills of creativity, complex problem-solving, and sustainability, which are increasingly critical in our rapidly evolving world. In this context, the article explores how integrating principles from the natural world into teaching methods can offer a new perspective. By drawing inspiration from nature's rich tapestry, educators can reimagine traditional teaching approaches, foster creativity, and promote sustainability.

One of the focal points of our discussion is the *SummerLab Biodesign* program, a renowned international initiative nestled within the enchanting Natural Park of Redes in Asturias, Spain. Through a series of meticulously designed learning modules, this program equips participants with a diverse array of methodologies, tools, and resources. Its primary goal is to cultivate a profound sense of curiosity and observation towards nature, ultimately fostering the discovery of alternative design processes through creative experimentation.

Furthermore, we investigate *Biodesign School*, a dynamic platform for families and children eager to explore the captivating realms of biodesign and biofabrication. Through a rich tapestry of engaging activities and hands-on experimental laboratories, young minds are introduced to the intricate interplay between creativity, science, and technology. These experiences occur within the direct and immersive backdrop of nature itself, emphasizing the importance of learning in harmony with the environment.

In each section, we will develop and delve deeper into each learning program according to the following concepts: reimagining pedagogy, lessons from nature, implementing nature-inspired learning, and, finally, impact and future directions. These concepts revolve around the pressing need to cultivate creativity, critical thinking, and sustainability in students. It will delve into how the processes and systems inherent in nature can serve as a source of inspiration for pedagogical innovation. Practical insights will be provided on seamlessly integrating nature-inspired learning into educational contexts. This exploration encompasses a rich set of examples and strategies designed for educators to easily adapt and implement into their teaching practices. Additionally, a meticulous assessment of the potential impact of nature-inspired learning on students and the broader educational ecosystem is presented. As we look to the horizon of education, this debate also sheds light on the future, delving into the directions that future research and implementation can take within the dynamic realm of educational environments.

In essence, this chapter underscores the paramount significance of utilizing nature as a source of inspiration and innovation. It advocates for a paradigm shift in education towards experiential learning, promoting a deep connection with the natural world. Through these programs, participants of all ages are empowered to think creatively, embrace interdisciplinary approaches, and contribute to the future of sustainable design and manufacturing.

7.2 SummerLab Biodesign

7.2.1 Nature as a Laboratory

Biomimetic Laboratory (https://laboratoriobiomimetico.com/) is a place that bridges Nature with design, science, and technology through education and research. Nestled in the Redes Natural Park, it serves as a learning hub for designing and manufacturing inspired by nature's processes.

The natural space of Redes (Fig. 7.1) is located in Asturias, in northern Spain. It is a protected natural environment that was declared a Natural Park in 1996 and a UNESCO Biosphere Reserve in 2001 [2, 3]. The area is characterized by lush forests of beech and oak trees, making it an ideal place to enjoy inspiring settings for relaxation, reading, drawing, or simply resting. Through the various activities we conduct in the Biomimetic Laboratory, we aim to promote knowledge of the park and its natural and cultural values. It's an open and wild laboratory where you can feel and experience the secrets and superpowers of nature.

At Biomimetic Laboratory, we connect the research efforts in biomimetics with classroom education, employing a creative and immersive methodology grounded in Nature. This approach enables individuals to learn how to translate biological inspiration into technical implementation. It fosters the ability to observe Nature with curiosity and engage in creative experimentation to discover alternatives within the design processes. Taking Nature as a model prompts us to reconsider and reshape the future of design through alternatives to conventional production methods.

Biomimetic Laboratory represents a unique institution that forges a vital connection between the natural world and the domains encompassing design, science, art,

Fig. 7.1 Natural Park of Redes

and technology, all while operating within the sphere of education and applied research. Our geographic location in Ladines, a picturesque mountain village nestled within the Sobrescobio region of Asturias, confers upon us the distinct privilege of complete immersion within an environment of extraordinary distinction: the Redes Natural Park and Biosphere Reserve, celebrated for its rich biodiversity and cultural heritage of profound significance.

Our fundamental commitment revolves around the development of learning methodologies deeply rooted in the principles of bioinspired design and the nurturing of sustainable, regenerative manufacturing practices. At the core of our mission is the creation of materials and processes that demonstrate unwavering environmental stewardship while simultaneously championing the principles of circularity within the domains of design and production. Through the lens of biomimicry, we tirelessly endeavor to draw insights from the natural world and harness the evolutionary strategies it offers. This pursuit enables us to facilitate the discovery of innovative, highly effective solutions.

For us, paramount importance lies in the unique context of our location. The direct connection with a wild and untamed environment defines our ethos. The outdoors, in essence, serves as our primary laboratory, providing an unrivaled platform for exploration and experimentation. Nature serves as our primary laboratory, a boundless source of inspiration and innovation. The rich ecosystems and intricate processes found in the natural world are a constant reminder of the vast reservoir of knowledge waiting to be tapped into. In our work at Biomimetic Laboratory, we recognize that the best teacher and model for sustainable design and manufacturing is nature itself.

By observing, studying, and learning from the natural world, we gain valuable insights into efficient and sustainable processes, ingenious solutions to complex problems, and elegant designs honed by millions of years of evolution [4–6]. This perspective informs our approach to biomimicry and inspires us to create materials and processes that align with the principles of ecological responsibility and circularity. In conclusion, Nature provides us with a blueprint for a more sustainable future [7–9]. It challenges us to reimagine how we design, make, and interact with the world around us.

7.2.2 An Immersive Learning Experience

In the realm of educational and outreach initiatives, the *SummerLab Biodesign* program, conducted internationally, stands as a prominent endeavor. Over the past three years, this program has been meticulously organized within the heart of the Redes Natural Park. Spanning a duration of seven days, the program offers participants a multifaceted experience through a series of comprehensive training modules.

The distinguishing feature of the *SummerLab Biodesign* experience is its deep integration with the natural environment. Participants are encouraged to engage actively with the surrounding biodiversity and immerse themselves in the rich cultural heritage of the region. This direct communion with nature serves as a catalyst for fostering a profound understanding of sustainable design principles and biomaterial innovation. The ultimate aim of *SummerLab Biodesign* is to empower individuals. It empowers them with the ability to create with unwavering respect for the environment. Nature's principles serve as the guiding light for regenerative and sustainable design practices, aligning perfectly with our commitment to ecological responsibility.

SummerLab Biodesign program represents an exceptional educational opportunity for participants to engage in immersive learning, develop practical competencies, and cultivate a deep appreciation for sustainable design guided by the principles of the natural world.

SummerLab Biomaterials

The summer course *SummerLab Biomaterials* was an immersive learning program that took place in the heart of nature, within the Redes Natural Park. Through various training modules, participants had the opportunity to discover how to manufacture materials of organic origin that were completely biodegradable. During the course, the applications of these biomaterials, which could be used as substitutes for textiles or plastics derived from fossil sources, were explored.

One of the key focuses of the program was learning to observe nature with curiosity and from a perspective of creative experimentation. This allowed participants to find alternatives in manufacturing processes, taking nature as a model and inspiration. This perspective led them to rethink and reshape the future of design by exploring more sustainable options in contrast to traditional production processes.

The *SummerLab Biomaterials* course was designed as an experience for both seasoned professionals and students who wanted to delve into the world of materiality. Throughout the program, participants experimented with the manufacturing of biomaterials, learned how to shape them, explored techniques for drying and preservation, and even engaged in research to enhance their properties.

The interdisciplinary and global approach was fundamental in *SummerLab Biomaterials*. This course was aimed at explorers who aspired to lead transformational projects, with a holistic vision that considered the principles of sustainable development. Students, professionals, and educators from various creative disciplines, such as design, architecture, product design, graphic design, engineering, jewelry, fashion design, and art, found in this program a unique opportunity to integrate new methodologies into the creation and execution of projects. It provided participants with an enriching and transformative learning experience where nature and creativity came together to explore innovative and sustainable solutions in the field of materials. Biofabrication is a field that merges scientific research and artistic creation, replacing industrial or mechanical systems with biological processes. It combines biology, technology, and design principles inspired by natural organisms, aiming to revolutionize current production processes and mitigate their significant environmental impact.

During the *SummerLab Biomaterials*, participants had the extraordinary opportunity to delve into the captivating realm of manufacturing and shaping biologically based materials. They also explored traditional craftsmanship and techniques that carry the wisdom accumulated over time. The program was thoughtfully structured into interconnected training modules, where the knowledge gained in each workshop was seamlessly applied to subsequent ones. This approach fostered creativity and empowered participants to develop projects in a more innovative and effective manner.

In summary, the *SummerLab Biomaterials* provided a unique experience where biofabrication, craftsmanship, and creativity converged to explore sustainable and cutting-edge solutions in material production. Below is a detailed breakdown of the course structure:

Capsule 1: Biomimetic Walk and Natural Pigments

On the first day of *SummerLab Biomaterials*, participants embarked on an exciting "biomimetic walk" that immersed them in the practice of biomimicry through direct observation of nature. This walk took place in the surroundings of the Redes Natural Park, a rich area in biodiversity and diverse ecosystems.

One of the most notable aspects of this day was the creation of the biomimetic "discovery notebook." This notebook proved to be an invaluable tool for all creatives, as it helped connect the bioinspiration offered by the natural world with the generation of innovative and sustainable designs. Through it, participants learned to observe and analyze nature in an entirely new way, with scientific rigor. They carefully recorded curiosities and outstanding features of biological organisms, which filled them with inspiration and creativity to build an inventory of natural shapes, processes, and ecosystems. In this way, each one could create their collection of Nature's discoveries, supported by precise observations.

During the biomimetic walk, guidelines were shared, and activities were carried out that were found in the notebook specifically designed for this experience. This notebook not only helped overcome the fear of the blank page but also provided a rigorous and scientific approach to the observation and documentation of nature.

The fundamental objective of this workshop was to discover bioinspiration as a starting point for generating possible design ideas, emphasizing the importance of detailed observation and scientific rigor. It was a day full of exploration and connection with nature, where participants learned to look at the world around them from a fresh and scientific perspective, ready to apply this knowledge to innovative and sustainable projects.

Additionally, during the nature walk, a workshop titled "*Color in nature: biological pigments and natural dyes*" was conducted. The purpose of this workshop was to explore the various processes through which nature generates colors and consider these ideas as sustainable alternatives in product design, especially in the industrial and textile fields.

It was explained that in nature, color can originate through biological pigments or as a result of specific structures, known as structural color. Furthermore, it was discussed how our perception of these colors depends on various factors, such as the chemical substances present in organisms and the interaction of light with them.

Participants had the opportunity to study fascinating examples of colors in the natural world. From butterflies with vibrant colors to intensely green leaves, multicolored lichens, iridescent insects, and unimaginably colorful flowers, different manifestations of color in nature were explored. Emphasis was placed on the fact that each hue serves a communication function, conveying information about the species, its behavioral patterns, or potential dangers through color codes manifested through interaction with light.

The main objective of the workshop was to teach participants how to extract natural dyes from nature and begin creating their own collection of sustainable dyes. This not only fostered an understanding of the natural processes of color generation but also provided them with the opportunity to apply this knowledge in the creation of environmentally friendly products. It was a day filled with scientific discoveries and creativity, where new possibilities for sustainable design were explored.

Capsules 2, 3, and 4: Biomaterials

The next few days, we immersed ourselves in the fascinating world of biofabrication, exploring how nature creates its structures, materials, and textures through detailed studies of organisms, using this knowledge as inspiration to develop sustainable alternatives in the production of new materials.

We began our exploration by delving into different case studies of organisms that serve as examples of bioinspiration for the creation of biomaterials [10–15]. We analyzed how nature has perfected the manufacturing of its structures over millions of years of evolution and how we can apply these principles to produce more sustainable materials. Participants studied the general concepts and fundamentals of preparing bioplastics [16], biotextiles [17, 18], and bioceramics. Then we put what we learned into practice step by step, experimenting with different recipes using various bio-based materials. It was a practical process that allowed us to deeply understand how biomaterials can be created from natural and renewable sources.

In addition to bioplastics, participants were introduced to experimenting with biotextiles and exploring potential applications with the aim of promoting circularity in fashion design. They studied different formulations and learned how to select the most suitable waste materials to create biotextiles. When considering the creation of a biotextile using agri-food waste, it's essential to focus on materials that, when processed, yield a higher amount of fiber. Examples include banana peels, chestnut shells, or corn husks. The bonding between these fibers enhances the fabric's strength, providing greater flexibility to fold it without breaking. Furthermore, participants had the opportunity to explore various techniques for working with biotextiles, such as folding, flat weaving, sewing, or laser cutting. This allowed them to experiment with material and design possibilities in the fashion sector.

Fig. 7.2 Bioceramics using mussel shells

Finally, a project centered around bioceramics and their application in product design, such as lighting fixtures, furniture, or even architectural cladding elements, was introduced with the aim of fostering innovation and circularity in these sectors. In contrast to traditional ceramic production processes, bioceramic materials (Fig. 7.2) do not require firing in a kiln and do not produce polluting volatile substances. Therefore, they represent a sustainable alternative for environmentally friendly production. Throughout this session, participants experimented with various techniques and biobases such as gelatin or starch. They explored options for agri-food waste materials and learned how to process them for use as additives in the creation of bioceramics. Participants had the opportunity to work on formulations on flat surfaces as well as with three-dimensional molds, experimenting with different textures and designs.

Capsule 5: Folding

The final training module explored the folding patterns found in nature and taught various techniques of modular origami. It began with a detailed analysis of different folding structures in nature, using unique case studies of plants and their adaptations to the environment as inspiration for design projects.

Participants acquired skills in paper-folding techniques, stimulating their creativity in crafting complex structures through origami. Origami was understood as a creative tool in the design process, applicable in various areas such as packaging design, decorative objects, furniture, or ephemeral architectural structures. In this workshop, paper became a versatile medium for creating three-dimensional objects using different folding techniques.

The main objective of this hands-on workshop was for participants to master the general concepts and fundamentals of folding, using various types of paper and cardboard (Fig. 7.3). Through this skill, they were able to develop prototypes,

Fig. 7.3 Folding workshop

decorative objects, packaging, or even pieces of art. It was an experience that combined manual dexterity with creativity in design, allowing participants to explore new dimensions in the world of folded paper.

Summerlab Biomimicry

Just like *SummerLab Biomaterials, SummerLab Biomimicry* is an intensive seven-day course that takes place entirely in the natural environment of Redes Park, located in Asturias, Spain. This program combines outdoor excursions and lessons with hands-on digital techniques in the workshop. The main objective of the *SummerLab Biomimicry* is to provide immersive training in the science of biomimicry.

During this educational experience, participants delve into the science of biomimicry, which involves learning from nature to apply that knowledge to design and problem-solving. Through various activities involving observation, design, and implementation, supplemented with outdoor dynamics in the heart of nature, students learn how to transform biological principles into innovative design. This Biomimicry summer course offers specific tools to help students develop nature-inspired design projects. These tools encompass technical, aesthetic, and conceptual skills that are essential in creating sustainable and efficient designs.

Throughout the week, participants have the opportunity to interact with experts in the field, including professionals and academics, who delve into the topics being taught. The curriculum covers the intersection of animal and plant physiology, design, and digital fabrication. Additionally, the value of learning from traditional crafts and techniques, which hold the accumulated wisdom of the ages, is emphasized. The training capsules that make up this program are designed to be interconnected. The knowledge gained in each one can be applied in subsequent capsules, fostering creativity in the projects developed by participants. These training capsules cover essential concepts of biomimicry, its applications, and working methodologies.

In conclusion, *SummerLab Biomimicry* offers a unique experience for both professionals and students looking to enhance their creative processes by acquiring skills to transform observations of nature into ideas that solve design problems. Furthermore, this program provides an opportunity to become familiar with digital and traditional manufacturing techniques for creating functional prototypes. Below is a detailed breakdown of the course structure:

Capsule 1: Biomimetic Walk and Bioinspired Design Methodology

In addition to the previously mentioned biomimetic walk, the first day of *Biomimicry SummerLab* begins with an introduction to an innovative bioinspired design methodology, using a "biomimetic canvas" as a key didactic tool.

The methodology presented, called "From Biology to Design," focuses on analyzing the adaptation strategies of natural organisms to their environments and how these strategies can be transferred to innovations in the field of design.

To understand how the principles of organism adaptation can be applied to create innovative designs, this methodology is approached from two main perspectives. The first focuses on nature and how to identify the strategies and mechanisms of adaptation used by biological organisms in various habitats. The second concentrates on design and how to abstract and transform ideas derived from nature into potential and effective solutions for fields such as architecture, fashion, products, art, and engineering. The proposed methodology consists of several phases of work, all detailed in the biomimetic canvas. These phases facilitate the transition from biological inspiration to technical implementation, providing a valuable tool for generating conceptual designs.

The primary goal of this methodology (Fig. 7.4) is to equip participants with specific skills and approaches for integrating biomimicry into their design process. By understanding how nature solves problems and adapts to its environment, participants can use these principles as a source of inspiration to create designs that are more efficient, sustainable, and adaptable to the changing conditions of the modern world. This methodology also fosters interaction between biology and design, promoting interdisciplinary collaboration and innovation in multiple fields.

Fig. 7.4 Bioinspired design methodology workshop in Nature

Capsule 2: Observation

The second training module aims to teach various observation techniques, including macro photography and the use of microscopes. These techniques are essential for exploring and studying nature and its organisms in detail. In the observation of nature, it is crucial to maintain an attentive and curious attitude, similar to that of a child. Often, we encounter elements that catch our attention, such as a peculiar leaf or an interesting texture on a tree's bark. Drawing these details can help us better understand nature, but sometimes it's difficult to stop and observe for as long as we'd like. This is where nature photography comes into play. This tool allows us to capture multiple records of organisms, taking our fieldwork into the studio, where we have more time and resources for in-depth analysis.

Macro photography focuses on capturing images of small objects or details in nature at a very close scale. It's a fascinating way to explore the natural world and reveal the beauty that often goes unnoticed. With this technique, we can examine flowers, insects, textures on tree bark, leaves, fungi, and much more in great detail. Special lenses and lighting techniques are used to capture the most precise details and vivid colors. Even with a mobile phone, we can achieve interesting results using specific tools and applications.

Microscopy, on the other hand, allows us to explore the microscopic world of nature. Through this technique, we discover a realm of details that our eyes cannot perceive directly. We observe the microscopic scale, where phenomena occur that influence the macroscopic processes we see in nature.

In this workshop, participants learn techniques for preparing and handling study objects, as well as processing, interpreting, and recording images of various natural organisms. During a field trip, they focus on natural objects and observe tiny details using magnifying glasses and portable digital microscopes. Later, in the workshop, they use a binocular loupe to examine specimens prepared on slides and explore different magnifications. This training module promotes a deeper understanding of nature by allowing us to explore both its macroscopic beauty and its amazing microscopic world.

Capsule 3: Biodesign

The main objective of the third training capsule is to initiate the transformation of biological inspiration and observation into concrete design ideas. This is achieved through the use of techniques such as artificial intelligence (AI)-assisted design and the definition of parameters and guided steps in the biomimetic canvas. In this phase, participants learn to apply digital and conceptual tools that allow them to translate lessons from nature into innovative and concrete solutions in fields such as architecture, fashion, product design, art, and engineering. The training capsule focuses on the practical application of biomimetics and provides participants with the skills necessary to bring their ideas to life and take them to the conceptual design stage.

In this context, the use of artificial intelligence (AI) in biomimetic conceptual design has great potential for rapidly and disruptively visualizing solutions. It is a tool that helps designers analyze and synthesize large amounts of data, providing a realistic image of a prototype. AI can also be used to model and simulate complex biological systems, which can help designers better understand how these systems work and how they can be applied in various design disciplines. It is important to note that AI should be used in combination with expert knowledge and design experience. AI cannot completely replace human creativity and design vision.

Capsule 4: How Nature Makes

The last training capsule has the main objective of implementing the designs generated using new digital manufacturing tools. In this phase, participants learn how to carry out their biomimetic design ideas using advanced digital manufacturing technologies. These technologies can include 3D printing, laser cutting, CNC milling, and other manufacturing techniques that enable the precise creation of prototypes and final products.

During this training capsule, participants acquire practical skills to translate their design concepts into physical objects (Fig. 7.5). They learn to use computer-aided design (CAD) software to create digital models of their designs and then use machines and digital manufacturing equipment to produce prototypes and real products. Furthermore, the possibilities of combining traditional materials and advanced new materials in the digital manufacturing process are explored. This allows designers to experiment with a wide range of design and functionality options.

In summary, this training capsule provides participants with the necessary skills to take their biomimetic design ideas from concept to physical reality using cutting-edge digital manufacturing technologies. It enables them to explore new ways of efficiently and precisely creating nature-inspired products.

Fig. 7.5 Prototypes of different bioinspired designs manufactured using 3D printing

7.3 Biodesign School

7.3.1 Reconnecting with Nature

Throughout history, humanity has maintained a close relationship with the natural environment that surrounds it. This connection is not coincidental but deeply rooted in the essence of our species, forged over thousands of years of evolution.

Like any living organism, humans are inherently linked to the natural cycles and processes that govern our planet. Our dependence on the atmosphere for the oxygen we breathe and on the land for food is a clear testament to this interdependence. Moreover, the influence of natural cycles, such as seasons and weather patterns, has left a profound mark on our evolution and the shaping of our societies.

However, in recent times, we have witnessed a growing distance and a profound disconnection from the natural environment that surrounds us. This phenomenon has led to a gradual detachment from natural cycles and rhythms in favor of a more artificial, synthetic, and accelerated lifestyle. This shift is characterized by industrialization, globalization, and the rapid growth of the world's population, largely driven by technological advancements. As a result, we have witnessed unsustainable exploitation of natural resources, once considered inexhaustible.

Research in fields such as ecology, biology, and environmental psychology has emphasized the importance of restoring the connection with nature in our daily lives. Renowned authors such as Edward Osborne Wilson [19], Richard Louv [20, 21], Rachel and Stephen Kaplan [22, 23], and Florence Williams [24], among others, have underscored the critical significance of the relationship between humans and nature. They have highlighted how this connection is not only essential for our survival in an increasingly urbanized and technological world but also for maintaining good physical and mental health.

Their research strongly supports the idea that a disconnection from nature can have negative consequences for people's quality of life and overall well-being. This is particularly relevant in the context of children, where the lack of contact with nature can result in a "nature deficit." As various authors like Rachel Carson [25], Catherine L'Ecuyer [26], and Katia Hueso [27] have pointed out, childhood is a crucial stage for cognitive, emotional, and physical development, and exposure to natural environments greatly enriches this process.

The disconnection from nature in childhood has become a cause for concern. Today's children spend more time in front of electronic screens than outdoors. This distancing from nature can lead to a range of issues, such as increased stress levels, reduced creativity, and decreased empathy for the natural environment.

It is imperative that we reconnect children with nature by providing them with meaningful outdoor experiences. This not only promotes their healthy development but also instills a deep love and respect for the natural world. In doing so, we are building a future generation that is more aware of the importance of preserving and protecting our planet.

Therefore, our relationship with nature is undeniably deep and rooted in our history as a species. However, in the modern era, we have experienced a growing disconnection that has led to a series of environmental and health problems. Reestablishing this connection, especially in childhood, is crucial to ensure a sustainable and healthy future for both present and future generations.

7.3.2 A Learning Program from Wonder

It is for all the reasons outlined above that we, at Biomimetic Laboratory, have recently launched *Biodesign School* as an innovative learning program with the fundamental aim of bringing children closer to the beauty and wonder of the natural world that surrounds them. Grounded in principles of wonder and enthusiasm, we immerse young individuals in an exciting journey of discovery that is firmly rooted in science and experimentation.

Biodesign School is not merely an educational program; it is a transformative experience. Through a scientific and experimental approach, children have the opportunity to explore the diversity of nature. They embark on outdoor adventures where they observe and learn from the local flora and fauna. They discover how nature has developed astonishing solutions to life's challenges, from the resilience of spiderwebs to the efficiency of bird wings.

Our program also fosters active experimentation. Children become budding scientists and designers, using the inspiration they draw from nature to address design and manufacturing problems. This not only stimulates their creativity but also equips them with practical skills that they can apply in the real world. *Biodesign School* aims to ignite the spark of curiosity and love for nature in every child. We believe that this early connection with nature is essential for their personal development and for the future of our planet. By cultivating a deep respect and understanding of nature from an early age, we are sowing the seeds for a future generation of environmental advocates.

Therefore, *Biodesign School* is conceived as an invitation to explore, learn, and create in the natural environment. It is an opportunity for children to discover the "superpowers" of nature and be inspired by them. We are committed to nurturing the next generation of leaders and stewards of our planet, and we believe that this initiative is a significant step in that direction.

The didactic programming of *Biodesign School* is in constant harmony with and carefully tailored to the changing seasons, allowing us to fully leverage the evolving natural environment of the Redes Natural Park or any other natural setting. Immersion in nature is a fundamental aspect of these sessions, which typically take place at the facilities of Biomimetic Laboratory.

Furthermore, throughout the year, *Biodesign School* program is conducted in collaboration with education departments from various institutions and foundations across Spain. This collaboration has enriched the teaching experience and expanded the reach of our educational program. Below, we describe some of the activities conducted throughout the year 2023, with more than one hundred participants during the spring and autumn editions.

Spring Edition

In this immersive spring edition, we celebrate the wonders of nature in this vibrant season, emphasizing the unique phenomena and benefits of connecting with nature during this time of the year.

Session 1: Imaginary Biomaterials

In this engaging workshop, families with children aged 4 to 12 are introduced to the world of biomaterials. The session begins with a nature walk, allowing participants to observe the surrounding natural organisms. This observation forms the foundation for the subsequent experimental activities conducted within the Laboratorio Biomimético. The primary objective is to connect the direct observation of natural organisms with scientific experimentation, creativity, and individual imagination. Participants will learn to fabricate bioplastics using a simple and experimental approach. This hands-on experience not only teaches the basics of biofabrication but also underscores the environmental revolution it represents. Families will

Fig. 7.6 Experimentation and play with bioplastics and other materials

engage in creating imaginative objects using biomaterials, exploring various shapes and textures.

This session adopts a learner-centered approach, encouraging children and their families to actively participate in the creative process. By linking nature observations with practical experimentation, it fosters a deep understanding of biomaterials (Fig. 7.6). The pedagogical approach emphasizes curiosity, exploration, and hands-on learning, ensuring that participants gain insights into the world of biomimicry.

In conclusion, participants in the "Imaginary biomaterials session" not only acquired practical knowledge about bioplastics and biofabrication processes but also delved into the realms of circular economy and regenerative design. They discovered how to harness the potential of agricultural and food waste as opportunities for innovative design. Through creative thinking and problem-solving, they not only invented and crafted imaginative biomaterial-based creations but also contributed to the broader concept of sustainable and circular material usage, aligning with the principles of regenerative economics. Furthermore, participants gained a profound understanding of supermaterials found in nature. These materials possess the remarkable ability to be manufactured and reabsorbed into the natural environment and ecosystem in a sustainable manner. This newfound knowledge empowers them to draw inspiration from these natural supermaterials, unlocking innovative possibilities for environmentally friendly design and production.

Session 2: The Tinyness of the Forest

This session invites families to explore the microcosmos of our environment. Equipped with magnifying glasses and microscopes, participants observe the incredible superpowers of tiny organisms that often go unnoticed. These observations serve as the foundation for the subsequent creative activity. The primary aim is to inspire awe and curiosity in participants about the microscopic world. Through observations, they will discover the intricate and often surprising features of tiny organisms. Following this, each family will design a project inspired by the micro-forest.

The pedagogical approach revolves around fostering a sense of wonder and curiosity. By introducing participants to the hidden wonders of nature, this session aims to instill an appreciation for the complexity and beauty of even the smallest life forms (Fig. 7.7).

In retrospect, the benefits of the "The tinyness of the forest session" were substantial. Participants cultivated enhanced observation skills, gaining a profound appreciation for the critical role of scale in design. They grasped the intricate interconnectedness of organisms and natural structures across hierarchical orders, drawing boundless inspiration from renowned references in biomimicry and ecological

Fig. 7.7 Practicing close observation of various natural organisms found in the forest

science. Moreover, this immersive experience fostered a profound and lasting connection with the natural world. It heightened ecological awareness and underscored the pivotal role played by microscopic organisms in ecosystems. These insights, acquired through direct engagement with the environment, have left an enduring impact on the participants' understanding of nature's intricate web.

Session 3: The Color Laboratory

This session explores the vibrant colors of spring in nature. Participants will learn that in the natural world, color originates from biological pigments or structural properties. Through various techniques, they will create natural dyes and use them to produce collages inspired by the colors of spring. The aim or objective is to deepen participants' understanding of color generation in nature and encourage creativity through the hands-on creation of natural dyes and art. Participants will gain insights into the ecological significance of color in various species.

This session was designed to stimulate sensory engagement combined with scientific exploration and creative expression. Participants were immersed in the fascinating science of color as it manifests in the natural world. Through hands-on activities, they not only learned about the principles of color in nature but also harnessed this knowledge to create their own artistic representations. This pedagogical approach engaged their senses, encouraged scientific inquiry, and fostered artistic creativity, resulting in a holistic and enriching learning experience.

As a result of this session, participants developed a comprehensive understanding of biomimetic concepts related to color, including structural coloration (Fig. 7.8). They learned sustainable ways to generate color without causing harm to the environment, exploring the use of natural dyes and the extraction of pigments from the surrounding natural world. This experience not only deepened their appreciation for the role of color in nature and its ecological significance but also enriched their artistic and creative skills as they created natural dye collages. They left the session with newfound knowledge and practical skills, aligning with the principles of biomimicry and sustainable design.

In summary, the spring sessions were meticulously designed to provide educational and inspirational experiences for children and their families, fostering a deep and lasting connection with the natural world during this enchanting season. These sessions not only imparted valuable knowledge and skills but also cultivated a profound bond with nature during the season of renewal. They epitomized the ethos of the *Biodesign School* program, seamlessly integrating education and inspiration, and empowering participants to become stewards of a more sustainable and harmonious world.

Fig. 7.8 Experimenting with natural pigments extracted from flowers, soils, and vegetables

Autumn Edition

The autumn edition of the *Biodesign School* program aims to harness the unique characteristics of this season to foster children's connection with nature. Autumn, with its display of colors and changes in the natural world, provides an ideal backdrop for exploring biodiversity and the mysteries of the natural world. During this edition, three workshops were conducted specifically for families with children aged 4–12. Each workshop had a duration of 2.5 h and focused on combining direct observation of natural organisms with experimental and creative activities.

Session 1: Observing Like a Snail

The main purpose of this first session was to stimulate detailed, microscopic-scale observation of the natural environment, an exercise that helps participants connect deeply with nature. Through the use of magnifying glasses and microscopes, children and their families had the opportunity to thoroughly explore small creatures and structures that often go unnoticed. This activity not only fostered an appreciation for the incredible details and characteristics of natural organisms but also encouraged scientific curiosity. Participants discovered a microscopic world full of surprises and marveled at the complexity of life on such a tiny scale.

The underlying pedagogy in this session focused on the combination of scientific observation and creativity. Children learned that science and art are not separate disciplines but can coexist and enrich each other. Carefully observing nature at a microscopic level provided them with a unique perspective that served as inspiration for the creative activity that followed. After exploring the natural microcosm, each family delved into creating sensory designs. This part of the workshop stimulated creativity by allowing participants to use what they had observed to bring their unique designs to life. The connection between scientific observation and artistic expression was evident as participants translated what they had discovered in nature into their own creations.

In summary, this session not only encouraged detailed observation and appreciation of microfauna and microflora but also demonstrated how science and creativity can harmoniously converge. This holistic approach not only expanded participants' knowledge of the natural world but also stimulated their curiosity and their ability to find inspiration in the beauty and complexity of life in all its forms and scales.

Session 2: How to Draw a Tree

The second session, titled "How to draw a tree," had several essential objectives. First, it aimed to teach participants to observe and understand the various forms that trees take in nature. This activity not only aimed to help children and their families appreciate the diversity in tree morphology but also to develop sharper observation skills. A highlight of this session was the introduction of the concept of fractals and their role in nature. Participants learned how fractals are patterns that repeat at different scales and how they manifest in the structure of trees and other natural elements. This understanding not only expanded their scientific knowledge but also allowed them to appreciate the beauty and elegance of fractal forms in nature.

The pedagogy used in this session was based on Bruno Munari's book *Disegnare un albero*, a prominent reference in the field of observing forms and designs in nature. Participants learned to observe and appreciate the various forms of trees in nature through practical exercises and drawing activities. This promoted the connection between direct observation and artistic representation, enabling them to explore how form and function intertwine in nature. The activity also helped them understand the richness and diversity of forms and functions in nature, a central concept in Jorge Wagensberg's book *La Rebelión de las Formas*. Participants not only learned to observe and draw trees but also discovered how these forms relate to specific functions in nature.

In summary, the "How to Draw a Tree" session aimed primarily to teach participants to observe and understand tree forms, introduce the concept of fractals, and promote collaboration and collective creation. This experience not only enriched their understanding of design in nature but also fostered their appreciation for the diversity in natural geometries and their ability to find inspiration in the complexity and harmony of nature.

Session 3: Autumn Notebook

The third and final session, titled "Autumn notebook" had several essential objectives. First and foremost, it aimed to teach participants to observe and record the characteristic details of the autumn season. This activity was intended to help children and their families develop a deeper appreciation for the sensory richness of this unique season.

The pedagogy employed in this session focused on the arrival of autumn and the variety of sensory experiences it offers. Families learned to observe and record the changing colors, leaf shapes, textures of tree bark, and the sensations that autumn provided. This detailed observation allowed them to connect more intimately with nature during this time of year.

Furthermore, participants created a discovery notebook in which they documented their observations, natural objects, color palettes, and personal experiences of autumn (Fig. 7.9). This activity fostered creativity and artistic expression, as each individual expressed their appreciation for the golden season in a unique and meaningful way. This experience not only enriched their understanding of autumn but

Fig. 7.9 Discovery notebook of the autumn

also strengthened their bond with nature and promoted an appreciation for the beauty and sensory richness of this season.

The autumn sessions offered a range of significant benefits for the participants. Firstly, they provided an opportunity to explore and appreciate the beauty and complexity of nature during a unique and ever-changing season. Participants developed skills in detailed and multisensory observation, allowing them to connect more deeply with the natural environment. Furthermore, these sessions fostered creativity and artistic expression through the creation of discovery notebooks. Participants had the freedom to capture their perceptions and experiences uniquely, promoting an appreciation for the subjective beauty of autumn.

From an educational perspective, these sessions helped participants understand important scientific concepts related to the autumn season, such as leaf fall and changes in nature. They also encouraged collaboration in group activities, as creating ephemeral art in nature involved working together as a team. Collectively, these sessions contributed to deepening the relationship between children and their families with the natural environment and cultivating a greater respect for the autumn season and its role in the cycle of life. These educational and sensory experiences strengthened the bond between participants and nature, promoting a deeper appreciation for the beauty and complexity of the natural world.

7.4 Conclusions

The utilization of nature as a model and a guide for educational processes yields profound benefits that resonate across various fields of study and practical applications. This chapter has explored the transformative potential of nature-inspired learning and design methodologies, emphasizing their advantages in revolutionizing traditional teaching approaches.

Biodesign, Biofabrication, and Biomimicry These innovative approaches draw from the intricate workings of nature to tackle complex challenges. By studying and emulating natural systems, we unlock new possibilities for sustainable problem-solving. These methodologies encourage a holistic understanding of the environment, fostering respect for the intricacies of the natural world.

Sustainability and Circular Economy Nature thrives on sustainability and circularity, with one organism's waste becoming another's resource. Adopting these principles in design and manufacturing minimizes environmental impact and promotes responsible consumption. Nature-inspired design inherently contributes to a more sustainable future.

Immersive Education Immersive learning experiences, such as those offered by programs like SummerLab Biodesign and Biodesign School, transcend traditional classroom boundaries. Hands-on interactions with the natural world deepen comprehension and foster a profound connection with the environment. This approach encourages curiosity, creativity, and critical thinking.

Interdisciplinary Collaboration Nature-inspired learning promotes collaboration across diverse fields. Scientists, designers, engineers, and artists come together to explore and apply nature's wisdom to various challenges. This interdisciplinary approach enhances innovation and opens doors to novel solutions.

Benefits of Nature-Inspired Learning The advantages of this methodology are multifaceted. It nurtures creativity, inspiring individuals to think beyond conventional boundaries. It instills a sense of responsibility toward the environment and encourages sustainable practices. Nature-inspired learning fosters adaptability, critical thinking, and problem-solving skills that are applicable in diverse contexts.

Scientific Rigor Nature-inspired learning maintains scientific rigor at its core. It involves thorough observation, analysis, and experimentation, aligning with the principles of empirical research. This ensures that the insights drawn from nature are not mere conjecture but are grounded in sound scientific understanding.

In conclusion, the integration of nature into learning processes offers a powerful and holistic approach. It empowers individuals to address complex challenges with innovative, sustainable, and environmentally responsible solutions. Furthermore, this methodology cultivates a profound appreciation for the intricate beauty of the natural world, encouraging the next generation to become stewards of our planet. Nature-inspired learning is not just an educational approach; it is a pathway to a brighter, more harmonious, and sustainable future for all.

References

1. United Nations (n.d.) Sustainable Development Goals. [Online] Available at: https://sdgs.un.org/goals. Accessed 25 Sept 2023
2. UNESCO (n.d.) Biosphere Reserves in Europe and North America. [Online] Available at: https://en.unesco.org/biosphere/eu-na/redes. Accessed 25 Sept 2023
3. Vázquez A, Díaz TE (2007) Parque Natural de Redes: Tierras de Bosques y Urogallos
4. Attenborough, D., & Hughes, J. (2022). A life on our planet: My witness statement and a vision for the future. Witness Books.
5. Attenborough D (2019) Life on Earth. HarperCollins Publishers. ISBN: 9780008294304
6. Dobzhansky T, Hecht MK, Steere WC (1968) On some fundamental concepts of evolutionary biology. In: Hecht MK, Steere WC (eds) Evolutionary biology, 2nd edn. Appleton-Century-Crofts, New York, pp 1–34
7. Benyus JM (2002) Biomimicry: innovation inspired by nature, 2nd edn. H. Collins Pub, New York
8. Lakhtakia A, Martín-Palma RJ (2013) Engineered biomimicry. Elsevier, Waltham. https://doi.org/10.1016/C2011-0-06814-X
9. Pohl G, Nachtigall W (2015) Biomimetics for architecture & design: nature – analogies – technology. https://doi.org/10.1007/978-3-319-19120-1
10. Barthlott W, Neinhuis C (1997) Purity of the sacred lotus, or escape from contamination in biological surfaces. Planta 202(1):1–8. https://doi.org/10.1007/s004250050096

11. Barthelat F (2010) Nacre from mollusk shells: a model for high-performance structural materials. Bioinspiration & Biomimetics 5(3):035001. https://doi.org/10.1088/1748-3182/5/3/035001
12. Bell EC, Gosline JM (1996) Mechanical design of mussel byssus: material yield enhances attachment strength. J Exp Biol 199(4):1005–1017. https://doi.org/10.1242/jeb.199.4.1005
13. Prum RO, Quinn T, Torres RH (2006) Anatomically diverse butterfly scales all produce structural colours by coherent scattering. J Exp Biol 209(4):748–765. https://doi.org/10.1242/jeb.02051
14. Vukusic P (2006) Structural colour in Lepidoptera. Curr Biol 16(16):R621–R623. https://doi.org/10.1016/j.cub.2006.07.040
15. He J, Wang Q, Sun J (2011) Can polar bear hairs absorb environmental energy? Therm Sci 15(3):911–913. Retrieved from https://doiserbia.nb.rs/img/doi/0354-9836/2011/0354-98361103911H.pdf
16. Rosenboom JG, Langer R, Traverso G (2022) Bioplastics for a circular economy. Nat Rev Mater 7:117–137. https://doi.org/10.1038/s41578-021-00407-8
17. Seidu RK, Eghan B, Acquaye R (2023) A review of circular fashion and bio-based materials in the fashion industry. Circ Econ Sustain. https://doi.org/10.1007/s43615-023-00303-z
18. Fuentes KM, Gómez M, Rebolledo H, Figueroa JM, Zamora P, Naranjo-Briceño L (2022) Nanomaterials in the future biotextile industry: a new cosmovision to obtain smart biotextiles. Front Nanotechnol 4:1056498. https://doi.org/10.3389/fnano.2022.1056498
19. Wilson EO (2021) Biofilia: El amor a la naturaleza o aquello que nos hace humanos
20. Louv, R. (2013). Last child in the woods: Saving our children from nature-deficit disorder. Atlantic Books Ltd.
21. Louv R (2011) The nature principle: reconnecting with life in a virtual age
22. Kaplan R, Kaplan S (1995) The restorative benefits of nature: toward an integrative framework
23. Kaplan R, Kaplan S (1989) The experience of nature: a psychological perspective
24. Williams F (2018) The nature fix: why nature makes us happier, healthier, and more creative
25. Carson R (1965) El Sentido del Asombro (The Sense of Wonder)
26. L'Ecuyer C (2012) Educar en el asombro
27. Hueso K (2021) Educar en la Naturaleza

Marlén López holds a PhD in architecture, specializing in the application of biomimicry in architectural design. She currently balances her work as an architect and designer at Volumínica Studio with the directorship of the education and scientific outreach department at the Biomimetic Laboratory in Ladines, located in the Redes Natural Park. With extensive experience and a passion for teaching, she has developed various educational programs and an innovative learning methodology centered around bioinspired design.

Manuel Persa After graduating as a graphic designer at the Higher School of Art of the Principality of Asturias (ESAPA) in 2007, he began his professional career in various studios as a graphic designer, but soon became interested in other artistic disciplines, which led him to continue training in Barcelona, where he studied the Master in Multimedia Design at the Escola Superior de Disseny i Enginyeria (ELISAVA) in 2009. There he began to experiment the materiality of graphic design, through multimedia and interaction technologies.

Between 2010 and 2018 he worked in different design studios as art director for multimedia and animation projects for advertising, video clips and events. Since then, he has been developing his professional experience in his own studio, Volumínica, specializing in brand design and conceptual strategy with a focus on avant-garde and sustainable solutions, at the same time that he has been creating a production line around digital crafts in Laboratorio Biomimetico.

With a noticeable maker and self-taught character, he has complemented his curiosity with training in various techniques such as origami for structural designs on paper, additive manufacturing or the creation of bioplastics, as well as traditional techniques such as construction with cebatu and basketry. He has been selected and participated in various design and sustainability programs such as "Circular Prototype" of the Biodiversity Foundation and Ecoembes; "Circular Weekend" by Enviroo and Gijón IMPULSA; and the "3D Printing Based on Nature Patterns" Classroom at the University of Oviedo thanks to which he began his research into the science of biomimicry.

In recent years he has worked around Nature as inspiration for his creative process, which has resulted in a work methodology that he has transferred to several teaching programs, with workshops specializing in new forms and manufacturing technologies. Also, lately he has carried experimentation and research with biomaterials to the field of art, participating in different exhibitions.

Chapter 8
Design of Bio-based Coated Textiles with Incorporation of Pine Wood Wastes That Mimics Leather

Ana Maria Lousada Soares Figueiredo, Leonardo Santana, Augusta Silva, Carla Silva, Jimena Alarcón Castro, and Jorge Lino Alves

Abstract The origin of raw materials has been a recurring theme in the industry, considering global warming and actions toward sustainable development. The process of (re)utilizing natural residues has been an area of increasing visibility. On the other hand, associations linked to animal rights have been pressing to inhibit the use of animal skin. The primary objective of this work is to develop sustainable coated textiles with pine wood waste for the creation of coated fabrics, employing biomimetic principles to attempt a material that holds the potential to serve as a viable alternative to traditional leather. The goal is to establish a value chain for wood waste, thus contributing to the circular economy paradigm. These new fabrics were submitted to different characterisation tests to evaluate their durability. An artefact that demonstrates the potential of the new material and mimics pinecones was designed and manufactured to demonstrate the viability of the new textiles.

Keywords Design · Material design · Coated textiles · Leather alternatives · Wood waste · Biomimetics

A. M. L. S. Figueiredo
Master in Product and Industrial Design, University of Porto, FEUP, Porto, Portugal

L. Santana
Department of Mechanical Engineering, UFSC, Florianópolis, Brazil

A. Silva · C. Silva
CITEVE, Braga, Portugal
e-mail: asilva@citeve.pt; cjsilva@citeve.pt

J. A. Castro
Universidad del Bío-Bío, Concepción, Chile

J. L. Alves (✉)
INEGI, Faculty of Engineering of University of Porto, Porto, Portugal
e-mail: falves@fe.up.pt

A. J. V. de Arruda, F. L. Palombini (eds.), *Biology, Biomimetics and Natural Design*, Environmental Footprints and Eco-design of Products and Processes,
https://doi.org/10.1007/978-3-031-65877-8_8

8.1 Introduction

This chapter was developed based on a dissertation submitted by the first author to obtain master's degree in Product and Industrial Design (MDIP) at the University of Porto. This project was accomplished through a collaboration between the above-mentioned University and the Technological Centre for the Textile and Clothing Industries (CITEVE). It was developed in the scope of the project be@t (https://bioeconomy-at-textiles.com), which is also a collaboration under the international project called "Diseño de una nueva arquitectura de materiales activos, vivos y adaptativos, baseados en principios de la biomimesis y la revalorización de residuos industriales del sector manufacturera de madera de Chile," originally from the Universidad del Bio—Bio in Chile. This international project proposed the development of "new active, live and adaptive materials" using principles of biomimetics and incorporating waste of pine wood that came from the wood industry in Chile to add value to this waste and reduce the environmental impact of this industry on the planet.

In this particular study, the main goal was to develop a more sustainable coated textile that incorporated wood waste with the objective of, through biomimetic principles, becoming a more sustainable alternative to traditional leather.

This study was developed in Portugal, and so, to enrich this experimental work, pine wood waste from the country was also tested. To demonstrate one of the many options the developed materials can be used for, a prototype of a fashion accessory, using these materials was developed.

8.2 Theoretical Context

8.2.1 Wood Waste in Chile and Portugal: The First Problem

Over the years, waste generation has become a great and very impactful problem. It is a global problem, and Chile and Portugal are no exception. In 2018 Chile generated a total of 19.6 tonnes of waste, according to the Chilean Ministry of Environment [7] while Portugal, between 2004 and 2020 increased its waste generation per capita from 452 kg to 513 kg [9]. As for wood waste, Chile generates approximately five million tons a year [10], while in Portugal the estimates are around 5.6 million cubic meters of waste per year [1].

Both countries have been making efforts to increase the value of wood waste, using it for purposes like wood pellets, wooden pallets, and energy sources, although these efforts to reuse waste haven't been enough. Therefore, it is crucial to promote initiatives to develop new materials that can increase the percentage of waste reused and its value, decreasing the waste generated and the industry's environmental impact.

8.2.2 *The Leather Industry: The Second Problem*

The leather industry is very polluting, and society has been getting more conscious and informed about it so, with this pressure and the actual environmental impact, there has been an increased interest in new alternative solutions to traditional leather.

Some manufacturers argue that the leather is already a by-product of the meat industry, claiming that it integrates circular economy principles, and for that reason, it is less impactful than it seems. However, due to the washings and treatments that the animal leather needs, it can be three times more environmentally impactful than the synthetic alternatives made from polyurethane (PU), which are not environmentally friendly [16]. Synthetic alternatives emerged as a cheaper substitute to leather since this material has always been expensive, so people are looking for a more accessible material that resembles leather. However, these alternatives are fossil-derived, they are non-biodegradable, and their manufacturing processes are also harmful to the environment in many aspects, such as water contamination, use of chemicals, and others [12].

As so, the search for leather alternatives that are less harmful and noxious to the environment is very important to reduce the impact of the leather industry, however, it is also true that it has been a struggle to mimic leather in all its characteristics and properties as it is a unique material with exceptional features.

The most recent and most researched alternatives have been the bio-based alternatives. To better understand what these are, they are "derived from materials of biological origin, excluding materials embedded in geological formations and/or fossilised" [8].

The advantages that these alternatives bring are quite a few, such as reducing the consumption of water during production, reducing energy consumption, less CO_2 emissions, and others [8] These alternatives are getting more and more available and there are already many different options made from or with the use of different bio-based sources such as, alternative leather made from pineapple fibres like Pinãtex®, *Desserto*, which is derived from cactus leaves, Fruitleather Rotterdam, which is made from mango waste, and the list goes on [12].

The processes to develop these alternatives are various and quite different from each other. These processes are not perfect, rather far from that; they still have some impact on the environment; however, they are a step forward towards the objective of finding better, less harmful, and more environmentally sustainable solutions [12].

8.3 The Project

As mentioned before, the objective of this study was the development of coated textiles that incorporated wood wastes and that, through biomimetic principles, could be a possible alternative to leather. Before presenting the results, it is relevant to explain what is biomimetics and the methods and steps followed to achieve the resulting materials.

8.3.1 Biomimetics

Biomimetics is a word that combines two others that come from the Greek 'bios', which means 'life', and 'mimesis', which means 'to mimic'; so biomimetics means to mimic what comes from nature or biology [4].

One notable example of the application of biomimetics is the flying mechanisms that Leonardo Da Vinci invented in 1488 by studying how birds could fly and trying to mimic them [4]. Nowadays, it is a strategy used in the development of new materials, products, or processes, mostly with the intent to get solutions more environmentally sustainable [3]. Biomimetics can be applied to any area of study, but considering specifically the textile industry, it has been applying biomimetic principles, for example, to leather replacement. The continuous research on leather properties and characteristics, and research to try and mimic these features in newly developed materials to obtain more sustainable solutions is a great example of biomimetics. [3] This study, within the scope of trying to find new solutions that can be leather alternatives, also involved biomimetics to obtain some of the leather features in these new materials.

8.4 Textile Coating

The textile coating industry seems like a rather recent one, however, indirectly it was born when people discovered that if they smeared animal grease on skin or a piece of cloth, it would be more resistant to water or liquids. Currently, it is an immense part of the textile industry and one that is in constant development. It is used for many purposes, including aesthetics, performance, and function, and is considered sustainable technology as it uses less water and energy and consumes fewer chemicals [15].

Many coating techniques vary between each other and serve different purposes. To achieve the two materials developed in this study, only one technique was experimented with, but many recipes of formulation were tried until the final results. The formulation is the combination of products necessary to form the fabric coating layer. It can be composed of a variety of products, depending on the purpose and the type of fabric being coated, but most commonly, the base is a polymeric matrix, and the other products are additives to adjust properties and other features of the mixture. Examples of these are thickeners, crosslinkers, antioxidants, emulsifiers, pigments, etc. [14]. In this study, although very similar, two formulations were made, one for each material developed. In the case of the material that incorporated the Chilean wood waste, the percentage of waste that was able to be incorporated was 10%. As for the base of the formulation, it was used a bio-based polymeric matrix that composed 89.7% of the mixture. The remaining 0.3% is the crosslinker, which was necessary to add to better incorporate the other components. It was also added post-mixing a thickener. This component is not accounted for in the percentages of the mixture, it is added drop to drop to thicken the paste. Finally, a topcoat was also added; however, this one was not added to the mixture, it was added as a layer on its own after all the layers of the formulation necessary were dried. As for the material that incorporated the Portuguese wood waste, it was only possible to incorporate 7%. The polymeric base used was the same as the other formulation but in this case, the percentage was 92.7%. The other 0.3% is attributed to crosslinker and that was also added to the thickener in the same conditions as in the formulation presented before. After the application of the necessary layers, it was also applied to the same topcoat, as mentioned in the formulation with the Chilean wood waste. Both pastes were mixed using a mechanical agitator. The first components to be combined were the polymer and the crosslinker, and then the waste was added slowly, while the paste was continuously mixed, one spoon at a time. The thickener was added last and as mentioned before, it was added drop by drop until it was visibly less liquid and sticking more to the spoon. To make sure the mixtures were ready to be used, the viscosities were measured using a viscometer. The pastes needed to be viscous enough so that they wouldn't transfer to the other side of the fabric, but not so thick that they wouldn't adhere to the fabric's surface.

Both wood wastes had to be prepared first to be ready to use in the formulations. As for the Portuguese wood waste, it was pine bark that came in its natural form and had to be dried first for 26 hours at 60 °C to remove humidity, avoiding lumps and aggregations on the paste. For it to be ready to incorporate in the paste, the waste had to be in a powder state with a granulometry of 0.2 mm, so it had to be ground, in four stages, using an industrial cutting mill, plus manual sieving to achieve that particle size. As for the Chilean wood waste, it was pine sawdust. It came in small particle sizes as it was already shredded wood, so it only needed two stages of grinding plus the manual sieving to achieve the necessary particle size. This waste didn't have any remaining humidity as it was already particles of pretreated wood; so, it did not need to be previously dried.

The pastes were applied through one of the coating methods, the knife on a roll. In this method, a knife is hung above the fabric being coated that also has a roll right below the knife, so when the knife is moved, the roll moves too (Fig. 8.1) [14, 15].

The height of the gap between the fabric and the knife can be adjusted to control layer thickness [14, 15]. More than one layer can be applied, which was the case, and each layer must be dried before the application of the next one. In the case of the material with pine bark waste, two layers each of 0.1 mm thickness were applied and in the material that incorporated the Chilean sawdust, it was necessary to have three layers with 0.1 mm of thickness each. The drying was done at 100 °C for 5 minutes, for each layer.

Before applying the coating formulation, the fabric must be prepared too. Sometimes it is necessary to perform some pre-treatments. In this project's case, many fabrics were experimented with, but in the end, a 100% recycled polyester

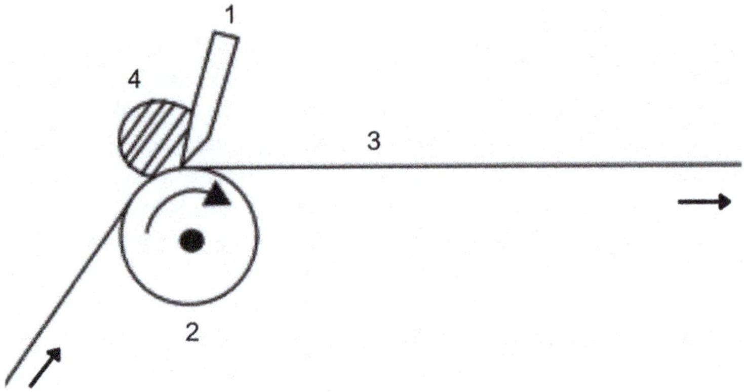

Fig. 8.1 Schematic diagram of the knife-on-roll coating method. (1) knife; (2) roll; (3) substrate; (4) coating formulation. Adapted from [13]

Fig. 8.2 Resulting materials incorporating the two wastes. (1) pine bark waste; (2) pine sawdust waste

fabric that was already ready to use was employed. After that, it was necessary to fasten it to the coating machine and stretch it to make sure it had no wrinkles, loose parts, or impurities on the surface, to ensure a smooth and homogeneous coating.

In the end, the coating is cured to seal and ensure it is finalised. In this study, the curing was done at 150 °C for 5 minutes. Some finalising processes can be applied with other products like a topcoat, which was one of the finishing processes used in this project, as mentioned before, and there were also mechanical processes like calendaring and embossing, which are processes that implicate heat and pressure. Calendaring was also applied in this project.

The results achieved for each experimental material developed were the ones presented in Fig. 8.2.

8.5 Materials Characterisation

Some analyses were made on both the formulations and the resulting materials to get preliminary information on the molecular structure, and thermal properties of the developed materials. The tests performed on the pastes were a Thermogravimetric Analysis (TGA), Differential Scanning Calorimetry (DSC), and Fourier Transform Infrared Spectroscopy (FTIR). As for the materials developed, the tests performed were colour fastness to washing, colour fastness to direct sunlight exposure, the resistance to abrasion Martindale surface test, the colour fastness to rubbing test, colour fastness to artificial light and a sewability test.

8.6 TGA

TGA is a test that provides information on the thermal stability of the material as well as on the start and end temperatures of degradation of each component present in the structure of the raw material under analysis. To do so, a small sample is heated and the weight changes that occur on the material are measured [5].

This test was performed on both designed materials formulations and on a "control" paste which was a formulation with the same ingredients except the wastes. The test was carried out on STA 449F5 equipment (Netzsch) in a synthetic air atmosphere for a temperature range of 28 °C to 700 °C, with a heating rate of 20 K/min.

Figure 8.3 shows 3 graphics each corresponding to one of the tested pastes.

In the graphic correspondent to the 'control paste', a plateau is visible and there is a decrease of around 5% of mass, which can indicate the release of a solvent or evaporation of water; however, this plateau does not exist in the other two graphics correspondent to the formulations with wastes. This can indicate some type of reaction between the wastes and other components that prevent that release.

Another observation is that the temperatures at which degradation begins (T_{onset}) tend to decrease slightly with the insertion of wood residues. These temperatures are low, below 100 °C, indicating the decomposition of some specific product of the reaction between the elements of the base pulp and the wood. However, it is observed that the insertion of residues improves the thermal stability of the pastes. If the overlap between the mass loss curve for the control paste and the formulation with wood residues from Portugal is observed, it appears that there is a smaller variation in mass, for the same temperature level, when there is the presence of bark pine (Fig. 8.4). This behaviour is favourable for the performance and thermal resistance of the products to be coated with the new mixture.

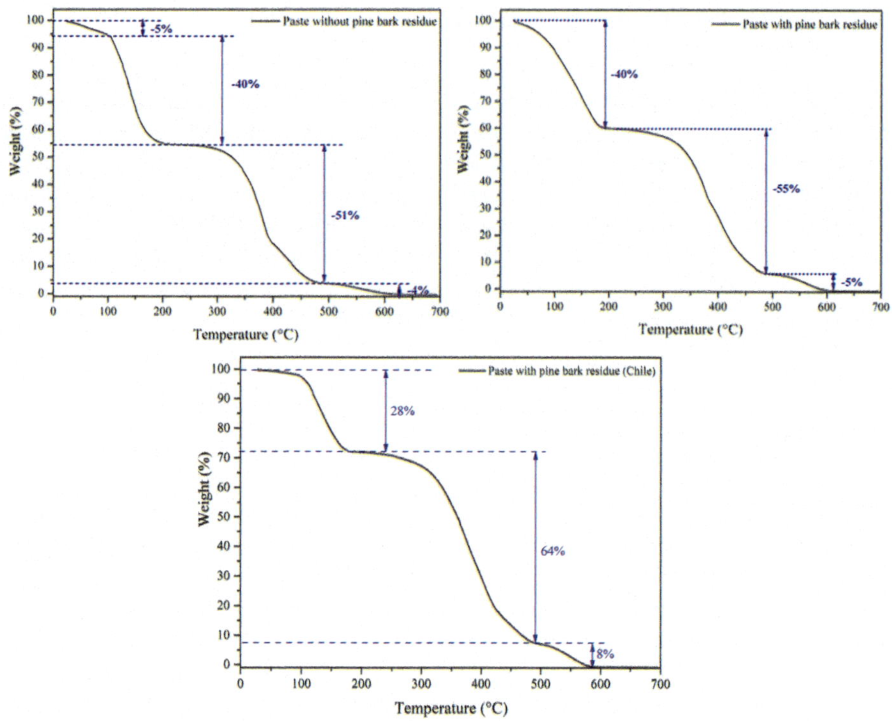

Fig. 8.3 Graphics with the results of the TGA tests

8.6.1 DSC

The DSC is a technique used to obtain information regarding the materials' thermal properties [2], in this case, the analysis was carried out to identify the effects of incorporating pine bark residues on the transition temperatures of the pastes in comparison to the control element. The test was conducted on a TA Instruments model Q20 equipment. The control paste and mixture samples were subjected to two heating cycles and one cooling cycle. The temperature range analysed was 50 °C to 180 °C with a heating rate of 20 °C/min.

The graphics in Fig. 8.5 show the results obtained in these DSC tests for the first heating cycle of each sample.

The information gathered was that the formulation without waste incorporated has a higher glass transition (T_g) and melting temperature (T_m) when compared to the formulation with pine bark waste. The reduction in the glass transition temperature (around −32%), with the insertion of residue, presents a negative result for the process since it increases the probability of the coating material becoming mobile

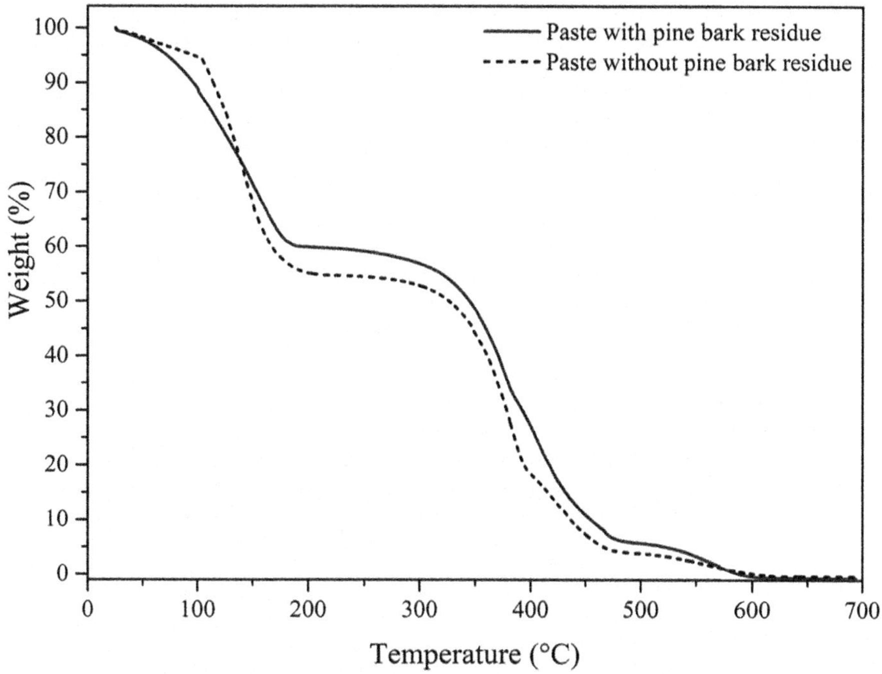

Fig. 8.4 Graphic that compares both materials' TGA results

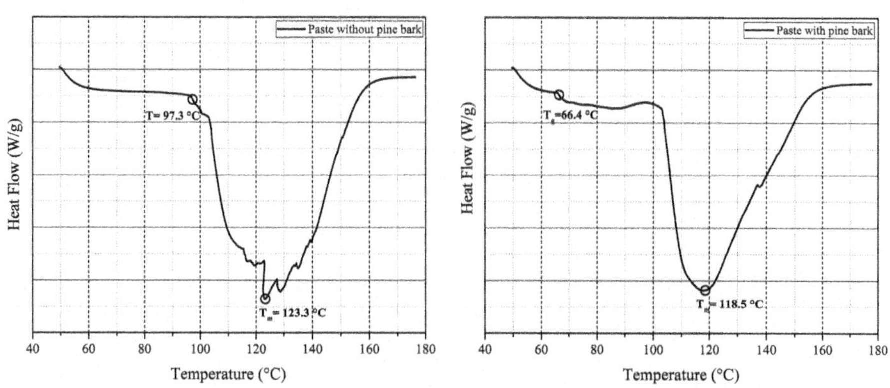

Fig. 8.5 Graphics that present the results of the DSC tests made

with the increase in the temperature of the operating environment of the coated products. Along with this information, it was observed that the formulation with the pine bark waste also presents the transition temperatures more evidently than the "control" paste.

With these results, it was possible to conclude that the formulation with pine bark waste has an advantage in having a reduced T_m because it can translate into a reduced use of energy required in production.

8.6.2 FTIR

Infrared spectroscopy (FTIR) was used, as an auxiliary technique, to identify possible phenomena (interactions, displacements, or disappearance of bands) in the molecular structure of the pastes, with and without pine bark, which could justify the alterations observed in the thermal behaviour of the paste's materials. A Cary 630 FTIR equipment, Agilent Technologies, was used for the analysis, with spectra being evaluated in the range of 4000 to 600 cm^{-1}.

The results obtained in this test are presented in the graphic from Fig. 8.6.

This test ended up not showing relevant results. As it is visible in the graphic, the waves from each paste tested, do not differ significantly from each other. It was concluded that the differences in the molecular structures of these formulations were not significant.

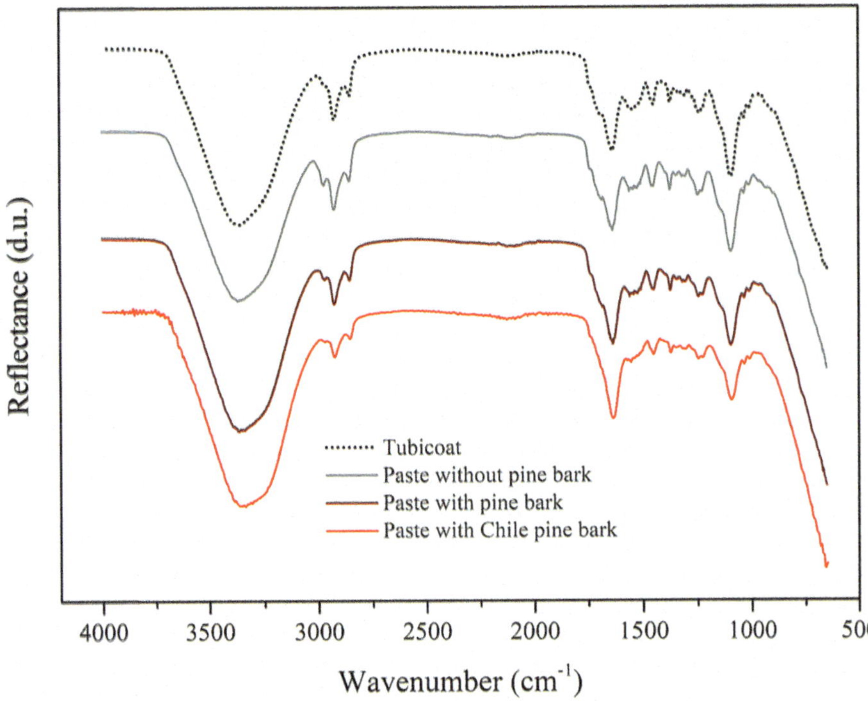

Fig. 8.6 Graphic with the results obtained in the FTIR tests

8.6.3 Colour Fastness to Washing

Now starting with the tests performed on the developed material's surfaces, the first one being presented is the test made to understand the materials' resistance to washing cycles. A sample of each material was exposed to none, one, five, ten and fifteen washing cycles made all in the same conditions, the same washing machine, using water and a common commercial laundry detergent.

Figure 8.7 shows a picture of the samples that underwent different numbers of washes, from each coated material.

These samples were then analysed in a Datacolor® Spectraflash® SF450® spectrophotometer, using an aperture of 6.6 mm and the UV filter UV D65/10 (Ganz-Griesser). The reading software was, Datacolor® Colortools®. This machine provided accurate colourimetric data and, in some cases, this data was compared to a set pattern. The results obtained are qualitative and quantitative. The quantitative results are provided in CIELAB colour space, K/S values and ΔE, the only quantitative result provided by comparing to the pattern. The qualitative results are all provided in comparison to the pattern and indicate if the sample being tested is brighter or darker, more or less saturated, and to which of the four colours used in CIELAB colour space (blue, green, yellow, red) it is closer to the pattern.

In this case, the pattern set was the sample of the material without any washes.

The results of this analysis are presented in Tables 8.1 and 8.2 below.

It was possible to conclude that, overall, there were some changes on the coating surfaces; however, not that significant in both materials. The biggest difference registered in the material with the pine bark waste colour surface occurred between the sample that was washed 10 times and the one that was washed 15 times, while in the material with pine sawdust, the biggest difference occurred between the sample washed 5 times and the one washed 10 times.

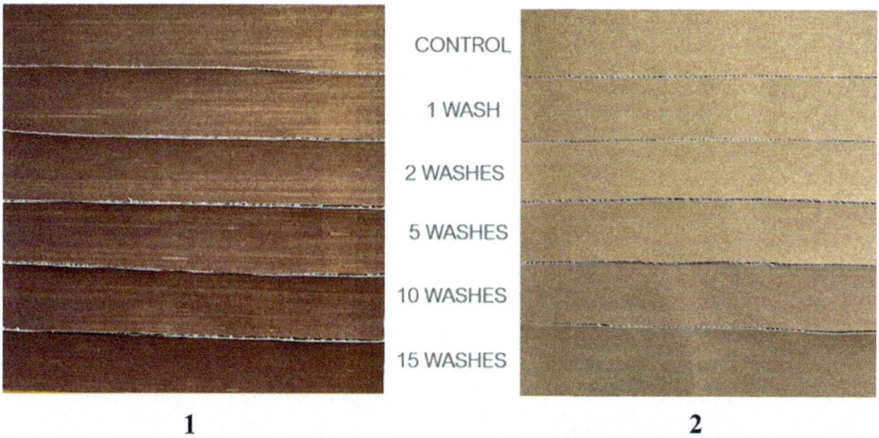

Fig. 8.7 Samples that were tested in washing cycles. (1) Material with pine bark waste; (2) material with pine sawdust waste

Table 8.1 Results of the analysis in the spectrophotometer obtained for the material with incorporation of pine bark waste that was exposed to washing cycles

Sample designation	L*	a*	b*	ΔE	KS	Qualitative results
Control pine bark (PB)	31.50	14.83	14.66	–	12.989	–
Washed 1× PB	31.32	14.80	14.38	0.34	12.660	Darker; less saturated, more red
Washed 2× PB	30.57	15.31	13.80	1.35	13.131	Darker; less saturated, more red
Washed 5× PB	29.30	15.26	13.65	2.46	14.631	Darker; less saturated, more red
Washed 10× PB	29.28	15.78	13.42	2.71	14.231	Darker; less saturated, more red
Washed 15× PB	28.50	14.36	10.79	4.92	13.303	Darker; less saturated, more red

Table 8.2 Results of the analysis in the spectrophotometer obtained for the material with incorporation of pine sawdust waste that was exposed to washing cycles

Sample designation	L*	a*	b*	ΔE	KS	Qualitative results
Control sawdust (Sd)	57.15	13.24	28.77	–	6.7429	
Washed 1× SD	57.33	13.03	27.37	1.43	6.2279	Lighter, less saturated, more red
Washed 2× SD	58.03	13.42	27.81	1.32	6.1475	Lighter, less saturated, more red
Washed 5× SD	57.24	13.31	27.20	1.57	6.2320	Lighter, less saturated, more red
Washed 10× SD	55.03	13.10	25.69	3.75	6.6429	Darker, less saturated, more red
Washed 15× SD	55.41	13.20	25.68	3.55	6.1161	Darker, less saturated, more red

Although there were slight changes on the coated surfaces between washes, both materials are considered to be resistant to washing, as the differences in the shades are not significant.

8.6.4 Sunlight Exposure Test

In this test, similarly to the previous one, a sample of each material was exposed to 0, 1, 2, 7, 14 and 21 days of sunlight exposure. To do so, a sample of each material was sectioned into 6 parts and set in a place with direct sunlight and progressively covered with black cardboard as the days went by so that the covered parts stopped being exposed to light.

Figure 8.8 shows a picture of the samples after different days of exposure to direct sunlight.

These samples were then analysed in the same spectrophotometer used in the fastness to washing test. The results obtained were presented using the same parameters and can be seen in Tables 8.3 and 8.4.

Contrary to what was observed with the washings test, the results of this test are more significant. Both materials present substantial changes in their surface shades as they are exposed to sunlight for an increasing period. The changes present differently on each material and occur in different periods of exposure but in all, both materials' shades get lighter when exposed to more days of direct sunlight.

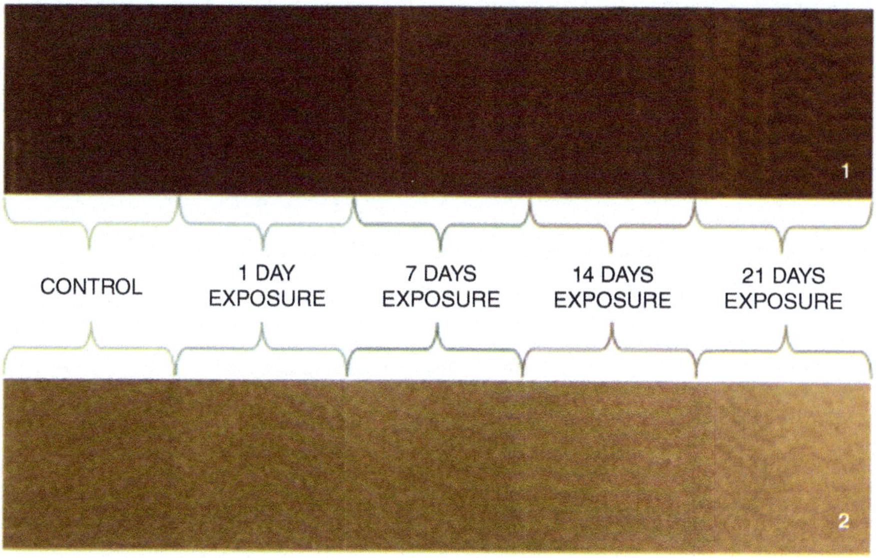

Fig. 8.8 Samples that were exposed to direct sunlight. (1) Sample with pine bark waste; (2) Sample with pine sawdust waste

Table 8.3 Results of the analysis in the spectrophotometer obtained for the material with incorporation of pine bark waste that was exposed to direct sunlight

Sample designation	L*	a*	b*	ΔE	KS	Qualitative results
Light control pine bark	32.17	15.20	14.66	–	12.342	–
Light 1 PB (1 day exposure)	31.04	16.03	14.31	1.45	12.906	Darker, more saturated, more red
Light 1 s PB (7 days exposure)	33.83	17.93	17.85	4.52	12.464	Lighter, more saturated, more yellow
Light 2 s PB (14 days exposure)	36.25	18.71	20.21	7.73	11.517	Lighter, more saturated, more yellow
Light 3 s PB (21 days exposure)	38.07	19.20	22.32	10.46	11.828	Lighter, more saturated, more yellow

It is possible to conclude that both materials could be improved regarding their resistance to sunlight exposure.

8.6.5 Martindale Abrasion Resistance Test

In this test, commonly used in the textile industry, a sample of the coated material is repeatedly rubbed against an abrasive surface to simulate daily life wear and tear.

Table 8.4 Results of the analysis in the spectrophotometer obtained for the material with incorporation of pine sawdust waste that was exposed to direct sunlight

Sample designation	L*	a*	b*	ΔE	KS	Qualitative results
Light control sawdust (Sd)	60.35	13.00	29.35	–	6.1796	–
Light 1 Sd (1 day exposure)	63.38	12.68	30.58	3.29	5.2490	Lighter, more saturated, more yellow
Light 1 s Sd (7 days Exposure)	63.56	12.72	32.63	4.60	5.7831	Lighter, more saturated, more yellow
Light 2 s Sd (14 days Exposure)	66.05	11.58	34.04	7.52	5.6437	Lighter, more saturated, more yellow
Light 3 s Sd (21 days Exposure)	68.47	10.66	34.81	10.07	5.3321	Lighter, more saturated, more yellow

Table 8.5 Results of the Martindale abrasion test on the samples with incorporated pine bark waste

Number of revolutions	Sample 1	Sample 2	Sample 3	Sample 4
Initial state	0	0	0	0
1600 revolutions	0	0	0	0
3200 revolutions	0	0	0	0
6400 revolutions	0	0	0	0
12,800 revolutions	1	1	1	1
25,600 revolutions	1	1	1	1
38,400 revolutions	2	2	2	2
51,200 revolutions	2	2	2	2

The conditions used in this test were:

- Type of abrasive: Wool fabric
- Applied pressure: 12 Kpa
- Used method: 1
- Conditioned environment: 20 ± 2 °C and 65 ± 4% H.R.
- Classification between 0 and 5 in which 0 is no alteration and 5 is complete degradation

The machine that performs this test records the number of cycles that are required to obtain visible wear or tear on the sample [17]. The results obtained are presented in Tables 8.5 and 8.6. More than one sample of each material is tested to have a more accurate result.

In the material with pine bark waste, it was possible to test four samples, while in the material with the Chilean sawdust waste, it was possible to test only two samples; but with these results, it was possible to gather conclusions. The material with pine bark waste starts showing signs of wear only at 12,800 revolutions and the material with pine sawdust waste shows signs of wear only at 38,2400 revolutions. These results indicate that both materials have very good abrasion resistance and when comparing both, the one with sawdust waste has a better resistance to abrasion than the material with the Portuguese pine bark waste.

Table 8.6 Results of the Martindale abrasion test on the samples incorporated with pine sawdust waste

Number of revolutions	Sample 1	Sample 2
Initial state	0	0
1600 revolutions	0	0
3200 revolutions	0	0
6400 revolutions	0	0
12,800 revolutions	0	0
25,600 revolutions	0	0
38,400 revolutions	1	1
51,200 revolutions	1	1

8.6.6 Colour Fastness to Rubbing

This test is performed to understand if the material when rubbed against another surface, transfers or stains. It can be performed using a dry and wet rubbing surface. Both designed materials were tested in both the wet and dry methods using a crock-meter, which is the machine that performs this test. The results are provided on a scale of 1 to 5, in which 5 is no colour transfer or staining and 1 is critical colour transfer/staining [11].

The conditions used in this test were the following:

- Used friction finger: round
- Applied force: 9 N
- Tested direction: web and weft or length and width
- Conditioning conditions: 4 hours at 20 ± 2 °C and 65 ± 4% H.R.

Both developed coated materials show a result of 4–5 on the scale when using the dry rubbing surface method, indicating that both are very resistant to rubbing and most likely won't transfer colour in dry conditions. As for the results with the wet method, the material with pine bark waste also presented a result of 4–5 while the material with pine sawdust waste got a result of 4. This means that when rubbed with a wet surface, the material with pine sawdust has a slighter chance to transfer/stain than the material with pine bark waste. However, these are still very good results, and it is possible to conclude that both materials are overall very good when it comes to resisting to not transfer or staining when rubbed. They most likely won't stain.

8.6.7 Colour Fastness to Artificial Light Exposure

This test determines how resistant a material is to artificial light exposure. Its focus is to determine if the textile being studied fades or discolours when exposed to this light [6]. The results of this test can be presented in many forms. In this case, CITEVE tested the developed materials using this technique in their labs and provided the results on a scale of 1 to 5 where 1 is very low colour fastness to light and 5 is very good.

The conditions for this test were:

- Used device: Xenon lamp air-cooled
- Used method: 3 (BS-5)
- Standard black thermometer temperature: 47 ± 3 ° C
- Humidity test: degree 5
- Blue scale assessment

The results obtained for the material with the Portuguese pine bark waste were classified with 3–4, while the material with the incorporation of the Chilean pine sawdust obtained the classification of 4–5. These outcomes indicate that the material with the sawdust resists better to light exposure than the material with pine bark. Similarly to what was concluded in the test of the colour fastness to direct sunlight exposure, both materials could be improved when it comes to their resistance to light exposure, independently of being artificial and natural.

8.6.8 Sewability Test

The last test performed on the materials was an informal sewability test. This test was made to understand if the material resisted well to being sewn in a sewing machine. To do so, a sample from both coated materials was sewed with 5 different types of seams. Figure 8.9 shows a picture of both samples tested.

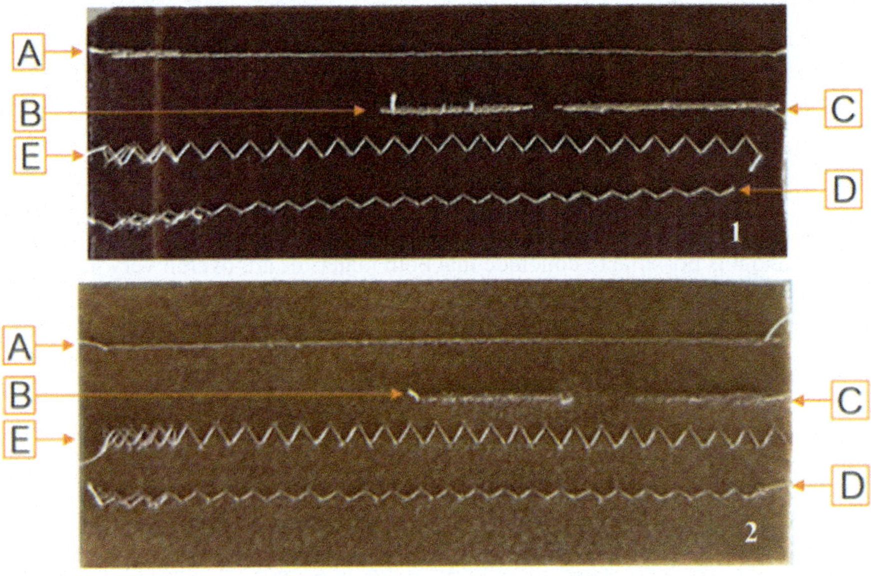

Fig. 8.9 Samples of each coated material with the sewing stitches. (1) Material with pine bark waste; (2) Material with pine sawdust waste

The results obtained were very good. None of the samples showed any breakage or significant tear in any of the seams performed.

8.7 The Showcase Object

To show an example of what can be done with the developed coated materials, an object was designed to display them.

As the developed materials are experimental leather alternatives, they eventually can be used for the same purposes as upholstery, decoration, fashion, or others, but due to some limitations on the amount and dimensions of the materials that can be produced at CITEVE, the options of types of objects that could exemplify possible applications for these coated materials were short.

The inspiration for the example object came from the tree from where the wastes that incorporate the developed materials are from, the Pine tree, more specifically, from the flower of said tree, the pinecone.

The final object is, then, a purse, a fashion accessory, inspired by pinecones. Another objective with this inspiration was to, in some way, highlight and "honour" the waste that is incorporated into the materials. Some initial sketches were made to explore the idea, and Fig. 8.10 shows some examples of these first sketches.

The main focus and features explored in this design were the overall cone shape and peculiarly shaped "petals" from the pinecone.

From the beginning of this idea, it was thought that the "petals" would be hanging from the base of the purse much like the real pinecones. Many designs for the shapes of the petals were tested in sketches and through mockups, but from the

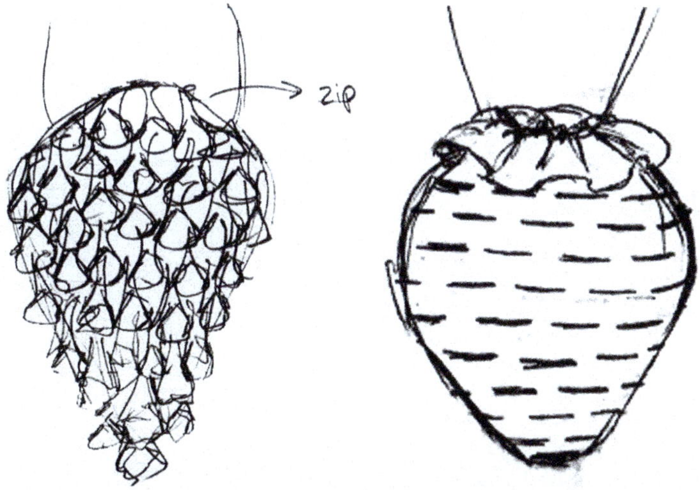

Fig. 8.10 Some of the first sketches for the proposed showcasing object

beginning, it was believed that to achieve a successful and clean-looking model, the shapes designed had to be synthesized/simplified, so this is what was made. In the end, the shape selected was the one shown in Fig. 8.11 below, and Fig. 8.12 shows a test petal in a sample of one of the materials, in the chosen shape.

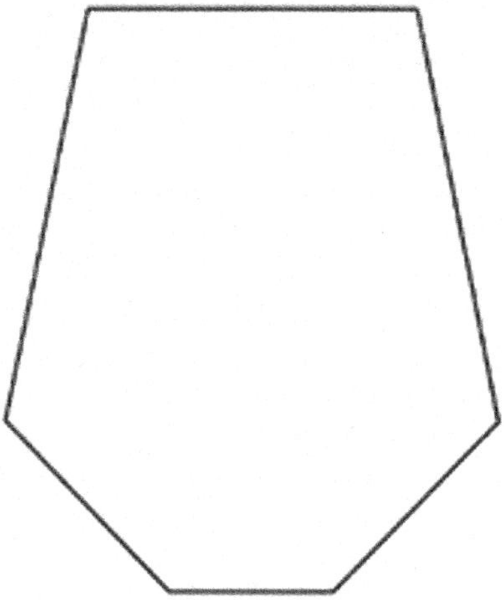

Fig. 8.11 Shape selected for the "petals"

Fig. 8.12 Mockup of the chosen shape using a sample of the chosen material

Fig. 8.13 Sketches of some of the ideas for the base shape of the purse

The base of the purse was also simplified and some shapes for it were thought of, though always having in mind the petals as they were the main focus. Figure 8.13 shows some of the designs that were thought of during the process of achieving the final piece and Fig. 8.14 shows the final chosen shape.

To produce the prototype of the purse, it was necessary to develop the sewing patterns of the pieces that would make the object. Figure 8.15 shows the pattern for the base of the purse. This pattern was used to cut two equal pieces with this shape using the material that incorporated Chilean sawdust to make the base for the purse (Fig. 8.16).

For the petals, the pattern developed is presented in Fig. 8.17 and it was necessary to cut 74 pieces of the material that incorporated the Portuguese pine bark waste with this shape to make the 32 petals that compose the purse as each petal needs two equally shaped pieces of the material (Fig. 8.18). These pieces of material that make the petals were cut with this shape in a laser machine (Fig. 8.19).

The last sewing pattern needed was the one to make the upper part of the purse (Fig. 8.20). Two equal pieces with this shape were cut from the material with the pine sawdust waste matching the base parts as it is visible in Fig. 8.21.

The final prototype is shown in Fig. 8.22.

It is relevant to highlight that whenever the purse is open, the petals are all close to each other, and much like one of the states, a pinecone can be found (Fig. 8.23).

However, when the purse is closed, due to the way it was designed, the petals "spike" up, to some extent resembling the state that a pinecone can be found in (Fig. 8.24).

This ends up giving the piece a sense of movement, dynamism and fun.

Although there were some constraints with material availability, these limitations were not considered relevant as it is believed that this purse has a strong and well-defined message that was possible to share through this type of object.

Fig. 8.14 Final sketch of the purse

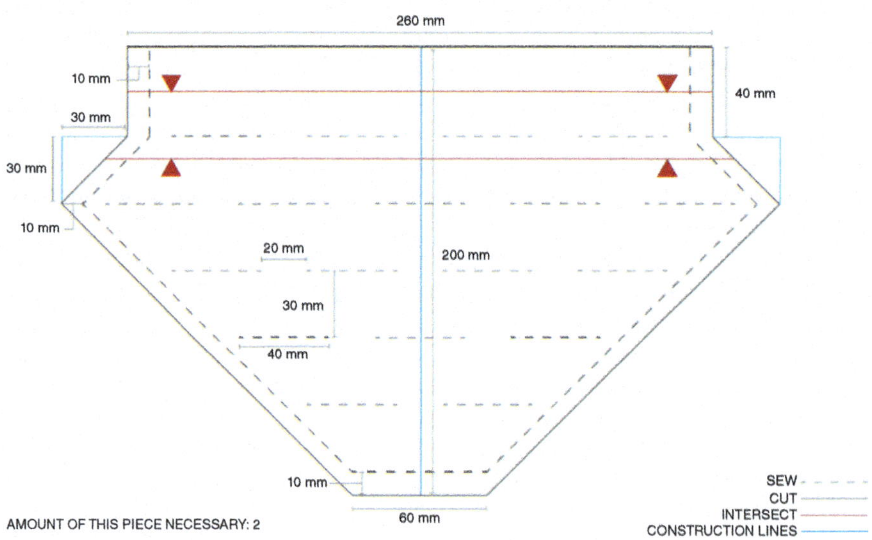

Fig. 8.15 Sewing pattern for the base part of the purse

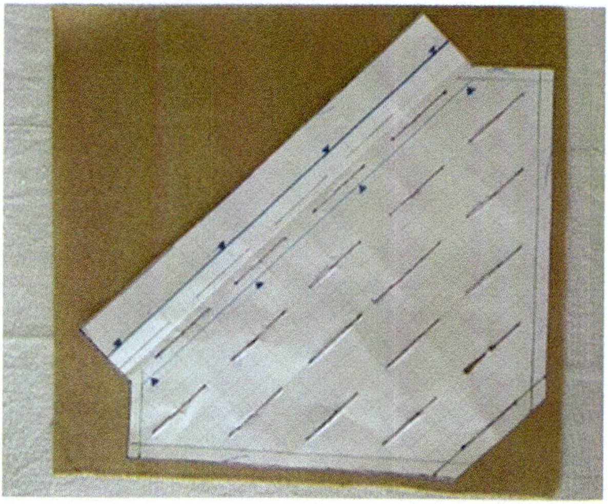

Fig. 8.16 Picture showing the moment of marking the pattern in the material

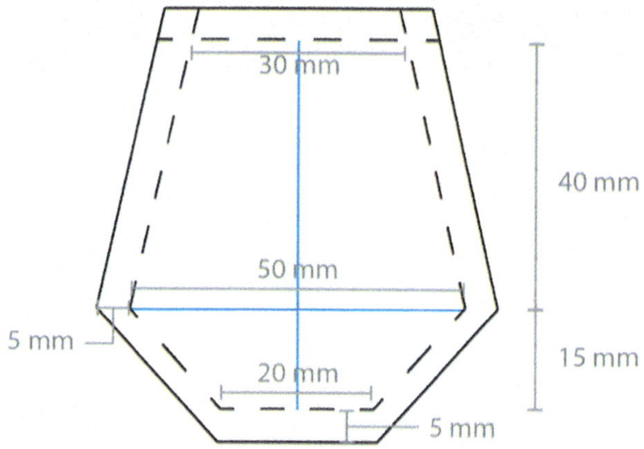

Fig. 8.17 Pattern created for the petals

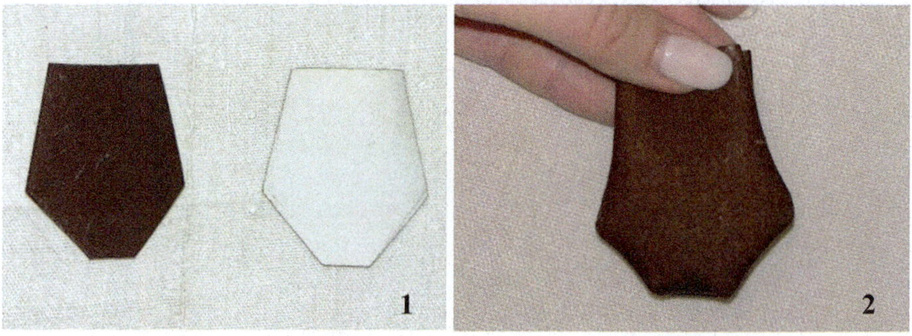

Fig. 8.18 The petal. (1) Petals pattern shape cut in the material developed; (2) a petal already sewn

Fig. 8.19 Picture taken in the moment of the laser cutting of the pieces that compose the petals

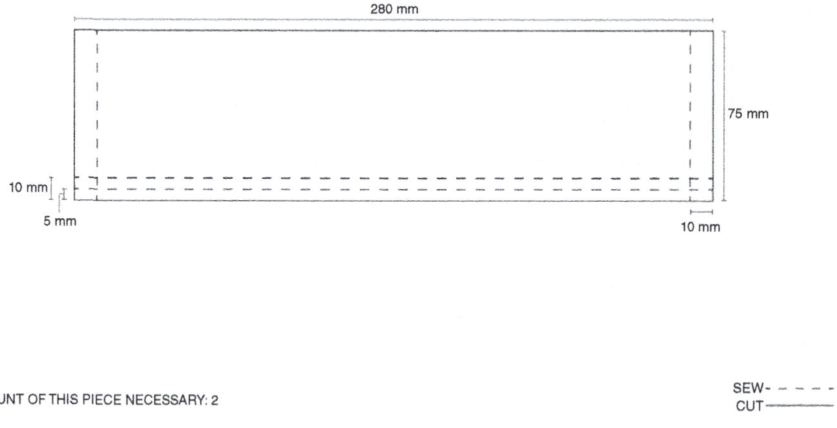

Fig. 8.20 Pattern developed for the upper part of the purse

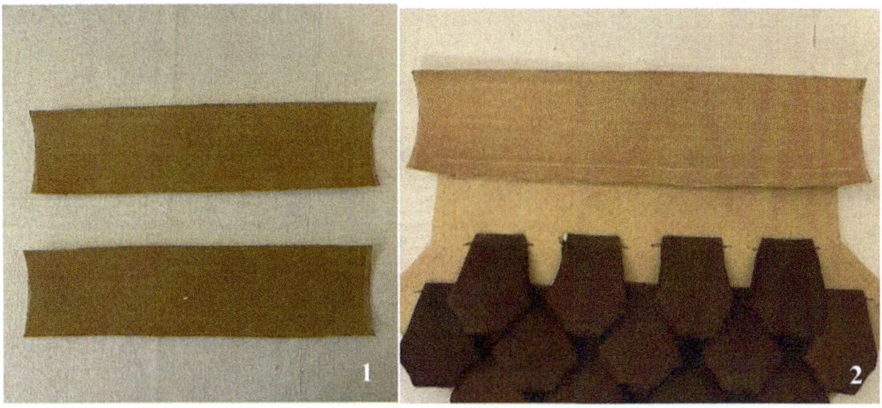

Fig. 8.21 The upper part of the purse. (1) Pieces of the material cut in the pattern shape; (2) one of the parts sewn to the base

Fig. 8.22 Final prototype, presented as being used

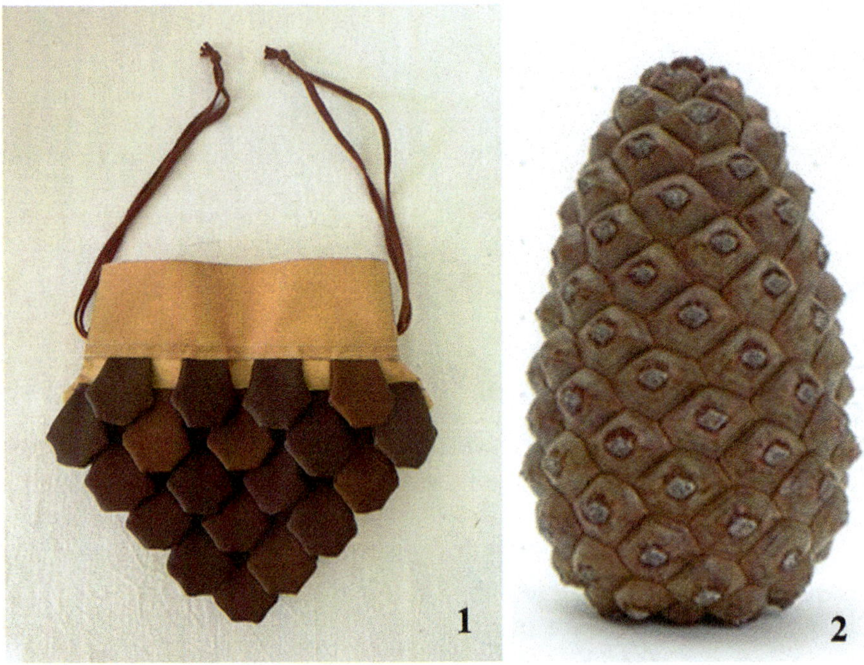

Fig. 8.23 Comparing the purse open (1) with a pinecone with closed petals (2)

Fig. 8.24 Comparing the purse closed (1) with a pinecone with opened petals (2)

8.8 Conclusions

To conclude, this study proposed the development of two coated textile experimental materials that could be viable alternatives to leather. These were indeed achieved by applying biomimetic principles, when considering the features of the touch and visual aspect that was desired for the surfaces of the materials, to be potential leather alternatives. Besides this, it was possible to utilize waste in these new materials, which can be considered a contribution to a more sustainable environment and use of resources, as well as a more conscious material development.

The possibility of incorporating the Portuguese pine bark waste and also developing a material using this waste did complement the study, providing a wider perspective on the matter and enhancing the work.

The designed purse and prototype ended up being a meaningful illustration of one of the many applications that the developed coated materials can have, exemplifying the value that wastes can have when well managed and the importance of establishing industrial symbiosis.

Acknowledgments The authors acknowledge the financial support from integrated project be@t—Textile Bioeconomy (TC-C12-i01, Sustainable Bioeconomy No. 02/C12-i01.01/2022), promoted by the Portuguese Recovery and Resilience Plan (RRP), Next Generation EU, for the period 2021—2026, and the project FONDECYT N° 1221361, financed by the National Research and Development Agency of Chile.

References

1. (IFN)., N. F. I. o. P (2010) Estimation of wood production and wood waste in Portugal
2. Ahmadi Khoshooei M, Fazlollahi F, Maham Y (2019) A review on the application of differential scanning calorimetry (DSC) to petroleum products. J Therm Anal Calorim:3455–3484
3. Barbosa de Araújo R, Arruda A, Alves J, Araújo P (2021) Biomimetic approach for experimentation and digital fabrication with bio-based material: the lightness and resistance of the Agave Sisalana. Cuadernos del Centro de Estudios de Diseño y Comunicación
4. Bhushan B (2009) Biomimetics: lessons from nature–an overview. Philos Trans R Soc A:1445–1486
5. Brown ME, Gallagher PK, Hwang JM (1998) Thermogravimetric analysis (TGA). In: Handbook of thermal analysis and calorimetry. Elsevier, pp 359–394
6. BSI Standards (2014) Part B02: colour fastness to artificial light: Xenon arc fading lamp test (ISO 105-B02:2014). BSI Standards
7. Cayumil R et al (2021) An overview on solid waste generation and management: current status in Chile. In: Environmental sustainability of current waste management practices, 21 October
8. Commission E (2023) Bio-based products. [Online]. https://single-market-economy.ec.europa.eu/sectors/biotechnology/bio-based-products_en
9. European Environment Agency (2023) Portugal Waste prevention country profile
10. Forestal CCN (2023) Waste generated in the forest sector. [Online]. https://www.conaf.cl/desarrollo-forestal/residuos-generados-en-el-sector-forestal/

11. NSAI Standards (2016) Textiles -tests forcolour fastness -Part X12: colour fastness to rubbing (ISO 105-X12:2016). NSAI Standards
12. Ribeiro HIC (2020) Criação de um acessório de moda em réplicas de couro sustentáveis. Design and Marketing of Textile Product, Apparel and Accessories
13. Sen AK (2008) Coated textiles: principles and applications. In: Coated textiles: principles and applications. CRC Press (Taylor & Francis Group), pp 69–97
14. Singha K (2012) A review on coating & lamination in textiles: processes and applications. Am J Polym Sci:39–49
15. Smith WC (2010) Smart textile coatings and laminates. Woodhead Publishing
16. Spinnewijn O (2022) Innovative leather alternatives: a short-lived trend or the future of fashion?
17. Wang X, Liu X, Deakin CH (2008) 4-Physical and mechanical testing of textiles. Fabric Test

Ana Maria Lousada Soares Figueiredo is a graduate student with a Master's degree in Industrial and Product Design from the Faculties of Engineering and Fine Arts at Porto University, Portugal. Her Master's thesis focused on the study of bio-based coated textiles, which led to the development of a book chapter in the field.

Leonardo Santana is a postdoctoral researcher in Materials' Science Engineering at the Federal University of Santa Catarina (Brazil). He holds a PhD degree in Mechanical Engineering from the Faculty of Engineering of Porto University (Portugal). His fields of research are nanocomposites made with thermosetting and thermoplastic polymers, mechatronic systems, and additive manufacturing technologies focused on extrusion-based open-source systems and statistical models.

Augusta Silva has a Master's degree in Textile Engineering. She is innovation manager for printing and textile coatings. She is a specialist and researcher in the departments of Chemistry and Biotechnology and Technology and Engineering at CITEVE, in the main areas of functional, advanced, technical, and intelligent textiles, with a focus on sustainability, circular economy, and industrial symbiosis, applied in the most diverse areas.

Carla Silva has a PhD in Textile Chemistry (University of Minho-School of Engineering). She is Director of the Department of Chemistry and Biotechnology at CITEVE. Carla research interests are in materials synthesis and functionalization, functional surfaces, environmental engineering and sustainability, bioengineering, and chemical engineering.

Jimena Alarcón Castro is a full professor at the Department of Art and Design Technologies of the Faculty of Architecture, Construction and Design of the Universidad del Bío-Bío, Chile. She has a PhD in Design Management (Universidad Politécnica de Valencia, Spain) and works in materials design with a sustainable approach, product design, and design management.

Jorge Lino Alves is a full professor at the Department of Mechanical Engineering of the Faculty of Engineering of the University of Porto, Portugal. He has a PhD in Materials Science and Engineering (Lehigh University, USA) and works in additive manufacturing, product development, and materials and processes characterization.

Index

A. J. V. de Arruda, F. L. Palombini (eds.), *Biology, Biomimetics and Natural Design*, Environmental Footprints and Eco-design of Products and Processes,
https://doi.org/10.1007/978-3-031-65877-8

The manufacturer's authorised representative in the EU is Springer Nature Customer Service Centre GmbH, Europaplatz 3, 69115 Heidelberg, Germany. If you have any concerns regarding our products, please contact ProductSafety@springernature.com

Printed and bound by CPI Group (UK) Ltd, Croydon, CR0 4YY
30/09/2026
02225471-0010